T0295315

Circular Economy for the Management of Operations

Mathematical Engineering, Manufacturing, and Management Sciences

Series Editor: Mangey Ram
Professor, Assistant Dean (International Affairs),
Department of Mathematics, Graphic Era University, Dehradun, India

The aim of this new book series is to publish the research studies and articles that bring up the latest development and research applied to mathematics and its applications in the manufacturing and management sciences areas. Mathematical tool and techniques are the strength of engineering sciences. They form the common foundation of all novel disciplines as engineering evolves and develops. The series will include a comprehensive range of applied mathematics and its application in engineering areas such as optimization techniques, mathematical modelling and simulation, stochastic processes and systems engineering, safety-critical system performance, system safety, system security, high assurance software architecture and design, mathematical modelling in environmental safety sciences, finite element methods, differential equations, reliability engineering, etc.

Circular Economy for the Management of Operations
Edited by Anil Kumar, Jose Arturo Garza-Reyes, and Syed Abdul Rehman Khan

Partial Differential Equations: An Introduction
Nita H. Shah and Mrudul Y. Jani

Linear Transformation
Examples and Solutions
Nita H. Shah and Urmila B. Chaudhari

Matrix and Determinant
Fundamentals and Applications
Nita H. Shah and Foram A. Thakkar

Non-Linear Programming
A Basic Introduction
Nita H. Shah and Poonam Prakash Mishra

For more information about this series, please visit: https://www.routledge.com/ Mathematical-Engineering-Manufacturing-and-Management-Sciences/book-series/ CRCMEMMS

Circular Economy for the Management of Operations

Edited by

*Anil Kumar, Jose Arturo Garza-Reyes,
and Syed Abdul Rehman Khan*

CRC Press
Taylor & Francis Group
Boca Raton London New York

CRC Press is an imprint of the
Taylor & Francis Group, an **informa** business

First edition published 2021

by CRC Press
6000 Broken Sound Parkway NW, Suite 300, Boca Raton, FL 33487-2742

and by CRC Press
2 Park Square, Milton Park, Abingdon, Oxon, OX14 4RN

© 2021 Taylor & Francis Group, LLC

CRC Press is an imprint of Taylor & Francis Group, LLC

The right of Anil Kumar, Jose Arturo Garza-Reyes and Syed Abdul Rehman Khan to be identified as the authors of the editorial material, and of the authors for their individual chapters, has been asserted in accordance with sections 77 and 78 of the Copyright, Designs and Patents Act 1988.

Library of Congress Cataloging-in-Publication Data

Names: Anil Kumar, K. (Anthropologist), editor. I Garza-Reyes, Jose Arturo, editor. I Khan, Syed Abdul Rehman, 1987- editor.
Title: Circular economy for the management of operations / edited by Anil Kumar, Jose Arturo Garza-Reyes, and Syed Abdul Rehman Khan.
Description: First edition. I Boca Raton, FL : CRC Press, 2021. I Series: Mathematical engineering, manufacturing, and management sciences I Includes bibliographical references and index.
Identifiers: LCCN 2020035629 (print) I LCCN 2020035630 (ebook) I ISBN 9780367422516 (hardback) I ISBN 9781003002482 (ebook)
Subjects: LCSH: Operations research. I Waste minimization. I Industries--Environmental aspects. I Social responsibility of business.
Classification: LCC TS57.6 .C525 2021 (print) I LCC TS57.6 (ebook) I DDC 658.4/034--dc23
LC record available at https://lccn.loc.gov/2020035629
LC ebook record available at https://lccn.loc.gov/2020035630

ISBN: 978-0-367-42251-6 (hbk)
ISBN: 978-1-003-00248-2 (ebk)

Typeset in Times LT Std
by KnowledgeWorks Global Ltd.

Contents

SECTION 1 Conceptual Understanding and Adoption Challenges of Circular Economy Practices

SECTION 2 Achieving Sustainability through Circular Economy Practices

SECTION 3   Applications of Advanced Methods in the Adoption of Circular Economy Practices

SECTION 4   Circular Economy and Related Area Practices in Operations Management

Preface

This book is divided into four sections. In the first section titled 'Conceptual Understanding and Adoption Challenges of Circular Economy Practices', which includes all the chapters related to the conceptual understanding and adoption challenges of circular economy (CE) practices. In the second section titled 'Achieving Sustainability through Circular Economy Practices' contains all the chapters related to achieving sustainability through CE practices. In the third section titled, 'Applications of Advanced Methods in the Adoption of Circular Economy Practices' contains all the chapters where the authors used advanced method in the adoption of CE practices. The last section is all about 'Circular Economy and Related Area Practices in Operations Management'. The brief description of each section as follows.

The section titled 'Conceptual Understanding and Adoption Challenges of Circular Economy Practices' contains three chapters. In the first Chapter 1, Priyanka Sihag, Aastha Dhoopar, Anil Kumar and Ashok Kumar Suhag discuss that there has been a continuous surge in the research around the concept of the CE. In today's world for the organizations to be effective, economic growth with minimum disruption for the environment has become essential. To address this issue, various organizations around the globe are switching to a CE wherein the natural resources are judiciously utilised and the wastage is minimised. In the CE literature, the 'human side of organizations' has seldom converged with the adoption of CE. This chapter aims to demonstrate how human resource management can contribute towards the adoption of CE. The relationship between human resources and CE focuses on the economic, social as well as the environmental dimensions, integrating the concepts of eco-innovation, leadership and top-management commitment, Green HRM practices and the supremacy of communication in the adoption of CE. The pivotal role of human resources in the transition towards the CE can be largely attributed to the stakeholders' theory and the resource-based view (RBV). The role of human resource management leads to developing a conceptual framework positioning the CE as a precursor to organizational sustainability. The recommendations for future research on the CE and the contribution of HRM towards the smooth transition to the CE are suggested.

To contribute the same section objective, in Chapter 2, Lucas López-Manuel, Fernando León-Mateos and Antonio Sartal talk that nowadays, society is becoming increasingly aware that there should be a new production and consumption model so the economy can internalize environmental and social impacts. As these concerns grow, industries face increasing stakeholder pressure to be transparent when reporting the environmental and social outcomes of their operations. In this view, CE initiatives should help minimize energy and raw material consumption while also being economically viable. The need for change and the challenges it entails are why the concept of circularity is obtaining collective attention in both the research community and organizations, particularly in the manufacturing sector. Our chapter aims to highlight the opportunities and difficulties of the Industry 4.0 context.

Through an analysis of Industry 4.0 technologies and their relationships, we describe the opportunities some of them offer (e.g. cloud computing, additive manufacturing, virtualization) to increase circularity in manufacturing processes.

In Chapter 3, Elaine Conway explains that over the last few years, there has been increasing interest in adopting new business models to respond to major environmental issues facing modern businesses, such as resource scarcity and rising energy costs. Many businesses are considering moving to CE-based business models, to reduce their dependence on these scarce resources and to support more sustainable business in the future. This transition can result in significantly different revenue streams, issues relating to asset valuations and investment requirements in comparison to a traditional linear business model that does not use CE principles. The impacts that these changes could have on the financial reports could be considerable, resulting in investor uncertainty. The accounting profession needs to consider how to mitigate these impacts and manage the new CE-based business models in conjunction with the firm's stakeholders. This chapter discusses the move to CE business models and their impacts on the financial reports. It also examines the potential role of integrated reporting (IR) in supporting CE business models, through its focus on value creation over the long term across a range of six capitals: financial, manufactured, intellectual, human, social and relationship, and natural.

Section 2 contains all the chapters related to achieving sustainability through CE practices. To follow this, in Chapter 4, Mahender Singh Kaswan, Rajeev Rathi, and Ammar Vakharia explain that in the modern era of high competition and climate risks, to remain competitive in the market, there is an immense need for clean technologies that not only enhance productivity but also reduces negative environmental impacts. Green manufacturing (GM) is an approach of sustainable development that improves the material and energy efficiencies and delivers high-quality eco-friendly products. This chapter outlines the grey areas of the GM right from the necessity to life cycle assessment, indicators of GM, development of GM system, and sustainable entrepreneurship. This chapter will facilitate the readers and practitioners to have a comprehensive understanding of sustainability in the system through the incorporation of the GM approach.

To follow the same section objective, in Chapter 5, Mudita Sinha, Leena N. Fukey discus that all the major industries are shifting from the linear economy to CE which aims at how the generated water can be transformed a value adds to the industry by modifying the production and consumption of the resources. The hospitality industry is also touched by this change. Hospitality and tourism industry has experienced an extraordinary growth in the recent decade that has catalyzed the requirement of the hotels which in turn has made people think about the serious issue of water generation which in the hotels. Kitchen, storage and lodging area of hotels are considered to generate a large amount of waste and hence makes it one of the primary sources of waste origination. So, this research focuses on the major difficulties and opportunities that hotels encounter by adopting this transformation.

In Chapter 6, Shikha Verma and Anukriti Dixit provide an extensive review of the recommended strategies and solutions presented in the existing literature while exploring the risks and opportunities associated with remanufacturing practices in the Indian automobile sector, the authors present a CE framework with potential pathways for innovation through the amalgamation of governance and industry practices. While the existing state of literature has explored barriers to and recommendations for more sustainable remanufacturing practices which focus on the quality of end products, the energy efficiency of the remanufacturing supply chain (SC) and rebates and relaxations offered through policymakers, this chapter presents a novel case for a cohesive and entirely cooperational systemic intervention, on the part of policymakers, industry partners and the consumers at large. While traditional business models focus purely on an inward-looking approach through the 'resources, skilled labour and revenue' trinity, we find that a more sustainable approach is through the development of cooperation in each leg of the 'state–consumer–industry' framework.

Section 3 entitled "Applications of Advanced Methods in the Adoption of Circular Economy Practices" contains all the chapters related to the applications of advanced methods in the adoption of circular economy practices. To support this in Chapter 7, Nurullah Umarusman discusses the change and development are an inevitable process in every environment where human beings are. Many factors such as technology, climate change, global politics and economy, and humans' desire to win directly affect this process. On the other hand, possible losses are wanted to be decreased, minimizing negativeness that these factors will bring. Struggle for economic, social and environmental development has always been at the forefront in the world's agenda through protecting scarce resources and managing them properly. Especially with the concept of sustainability, companies have played an important role in this process, and sustainability has spread to members in the SC. One of the most significant elements ensuring sustainability to be implemented in the chain is the supplier selection process. In this process, methods have an impact on procuring sustainability besides the criteria to be used in supplier selection. In this study, Fuzzy Global Criterion Method (Fuzzy GCM) has been used to solve the Sustainable Supplier Selection Problem (S-SSP) of a business operating in the construction sector. In the Fuzzy GCM algorithm, which includes the GCM, Compromise Programming (CP) and Fuzzy Linear Programming approach, these three approaches contribute to the solution of the problem at different stages. For this reason, besides the final solution provided by the algorithm, the results of the three methods make it possible to make different comments. Besides, in this study, three different classifications have been made in terms of the criteria that constitute the objective functions in Multiple Objective Decision Making (MODM)/Fuzzy MODM methods, MODM/Fuzzy MODM methods used and supplier selection type.

Contribute to the same section objective, in Chapter 8, Mukesh Kumar, Rahul S. Mor, Sarbjit Singh, Vikas Kumar Choubey explain that today, the manufacturing sector necessitates a highly productive system, maintenance-free machinery and multi-skilled operator for enhanced sustainability and productivity gains. The total productive maintenance (TPM) can serve as a means of attaining these goals as well as increased productivity for industries. In this study, TPM is implemented in the manufacturing industry and overall equipment effectiveness (OEE) is calculated for machinery performance. The data is collected through a questionnaire study conducted for the employees as well as breakdown summary sheets of the industry. In the analysis, it has been observed that the forging machine takes a long time to set up, and has a maximum number of non-value-added movements in the process. A single minute exchange of die (SMED) tool is implemented to reduce the setup time of the forging machine to 67 minutes per setup which in turn allows production of approximately 984 more products per day. A sustainable maintenance schedule is implemented for better performance of machinery and to train the workers to detect faults in the machine. A comparison of OEE before and after TPM implementation is carried out and found a significant improvement, and hence concluding that TPM helps the industry to achieve sustainability in the manufacturing industry.

Chapter 9 is all about using a Multi-Criteria Decision-Making Model for agricultural machinery selection in which, Ali Jahan, Alireza Panahandeh, Hadi Lal Ghorbani discuss the proper selection of mechanical harvesters, such as commercial collectors and cleaners, based on the crop conditions can increase the efficiency. The objective of this study was to select the most optimal mechanism for machine harvesting of cotton using an integrated model. For this purpose, three cotton harvester indicators were considered including harvest efficiency, trash bin content and Gin turn out. Due to data uncertainty in the harvesting mechanism section, interval data were used to represent the harvesting mechanism score. Based on the methodology of the study, SIMOS and ELECTRE-IDAT methods were combined. Therefore, the revised SIMOS method was used to weight the criteria and ELECTRE-IDAT method was employed for ranking the available options. Finally, using BORDA method, the sensitivity and validity of the study were analyzed. According to the obtained results, the spindle harvesting mechanism was selected as the best mechanism for harvesting cotton. The next priorities were brush, paddle and finger harvest mechanisms. Such methodology is useful for both buyers of cotton harvesters and farmers.

Section 4 contains all chapters related to CE and related area practices in Operations Management. In chapter 10, Arvind Upadhyay and Jeagyung Seong examine the ways small and medium-sized enterprises make use of green supply chain management (GSCM). It employs qualitative methods to examine GSCM with a case study and approaches used to include in-depth interviews and document analysis. This paper has identified a gap in the existing business and management research literature regarding small and medium-sized enterprise (SME) use of GSCM and has attempted to fill this gap. The research conducted finds that tactical and structural change can result in more environmentally friendly practices in SMEs. As confirmed by the case study, the structural change necessary for instituting GSCM involves a range of factors including innovation competency, cost savings, managerial arrangement, human resources and competitive advantage. However, it has also become evident that additional research is necessary to fully determine the ways management practice can impact SME sustainability.

To contribute the same section objective, in Chapter 11, R. A. Dakshina Murthy and Leena James discuss CE practices in the retail sector. They explained that the retail sector has seen tremendous growth and is expected to grow @ 15–20% in the next decade. The core of the retail operation is the Logistics Operation, which ensures that the material required at the point of sale is available when it is most needed. The adverse effects of Logistics operation result in the degradation of the environment due to emissions of Green House Gases (GHG) into the atmosphere. Through the adoption of Green Logistics detrimental effects of Logistics on the environment can be minimized. The research objective here was to ascertain the level of awareness about Green Logistics among the organized retail players and review the operational level practices of Green Logistics by the retail sector leading to identifying the main drivers for adoption of Green Logistics practices to achieve a CE. The population included retail players dealing with categories such as Food and Grocery, General Goods, Apparel and Consumer Electronics. Government Policy and Regulations, Environmental Degradation and the CSR & Corporate Leadership in Sustainability have, significant influence on Sustained Adoption of Green Logistics and the results of ANOVA indicate that for all the three constructs, the respondent's views do not differ between one another among various categories of retail. Measures to achieve CE through Green Logistics by organized retail units to establish sustainable operational practices have been suggested.

In Chapter 12, Sarika Yadav, Rahul S. Mor and Simon Peter Nadeem explain how CE practices can achieve in the agriculture sector. They mention that agriculture is a major source of livelihood for the majority of the Indian population, but the declining role of agriculture in the country's GDP, inflation and failing food security and sustainability policies are the concerns currently. The contribution of the food sector in net carbon emission is a key issue for policymakers who for long have relied upon 'Green Revolution' as an answer to feeding the country. The flaws in agri-food policies are now being exhibited in record-breaking inflation, food import and changing climate patterns. To solve the aspects of India's repercussion of our disfigured stratagem, quick actions are imperative in terms of adapting food models that would suffice the current and future demands of food supply without tempering the adequate climate pattern. 'Holistic Mitigation and Adaptation' is the phrase that stands for 'holistically mitigating current ecological damage and adapting quick food system models to sustain the availability of food to all without disturbing the ecosystem. This chapter represents the current scenario of the Indian food system, the loopholes in the agri-food policies, the effect on the environment, and a sustainable food system model based on an array of micro-entrepreneurship that mitigates the current damage and adapts sustainable steps towards the re-establishment of food security.

The last chapter of this section is all about Supply Chain Network design models for a CE: A Review and a case study assessment. This chapter, Sreejith Balasubramanian; Vinaya Shukla; Arvind Upadhyay, Mahshad Gharehdash and Mahnoush Gharehdash explain that the global SCs are getting increasingly dispersed, and hence, more complex. This has also made them more vulnerable to disruptions and risks. As a result, there is a constant need to reconfigure/redesign them to ensure competitiveness. However, the relevant aspects/facets for doing so are fragmented and scattered across the literature. This study reviews the literature to develop a holistic understanding of the key considerations (environment, cost, efficiency and risks) in designing/redesigning global SCs. This understanding is then applied to assess the global SC network of a leading multinational tire manufacturing firm; also, to provide recommendations on redesigning it. The study has significant practical and research implications for global SC management.

Thanks
On behalf of the Editors

Acknowledgements

The editors would like to acknowledge the help of all the people involved in this project and, more specifically, to the authors and reviewers that took part in the review process. Without their support, this book would not have become a reality.

We would like to thank each one of the authors for their contributions. The editors wish to acknowledge the valuable contributions of the reviewers regarding the improvement of quality, coherence, and content presentation of chapters. Most of the authors also served as referees; we highly appreciate their double task.

We are grateful to all members of CRC Press: Taylor & Francis Group publishing house for their assistance and timely motivation in producing this volume.

We hope the readers will share our excitement with this important scientific contribution the body of knowledge about various applications of CE for the Management of Operations.

Dr. Anil Kumar
Guildhall School of Business and Law,
London Metropolitan University, London, UK

Prof. Jose Arturo Garza-Reyes
Head of the Centre for Supply Chain Improvement
College of Business, Law and Social Sciences
The University of Derby, UK

Dr. Syed Abdul Rehman Khan
School of Economics and Management,
Tsinghua University, Beijing, China

Editor biographies

Anil Kumar is a Senior Lecturer (Associate Professor) at Guildhall School of Business and Law, London Metropolitan University, London, UK. For the last ten years, he has been associated with teaching and research. Before joining London Metropolitan University, he was Post-Doctoral Research Fellow in area of Decision Sciences at Centre for Supply Chain Improvement, University of Derby, United Kingdom (UK). He earned his PhD in Management Science from ABV-Indian Institute of Information Technology and Management, Gwalior, India. He did his graduation in Mathematics (Hons) and MSc (Mathematics) from Kurukshetra University, India. He earned his Master of Business Administration (MBA) and qualified National Eligibility Test (NET), June 2011. He has contributed over 60+ research papers in international referred and national journals, and conferences at the international and national level. He has sound analytical capabilities to handle commercial consultancy projects and to deliver business improvement projects. He has skills and expertise of Advance Statistics Models, Multivariate Analysis, Multi-Criteria Decision Making, Fuzzy Theory, Fuzzy Optimization, Fuzzy Multi-Criteria Decision Making, Grey Theory and Analysis, Application of Soft-Computing, Econometrics Models etc. His areas of research are sustainability science, green/sustainable supply chain management, customer retention, green purchasing behaviour, sustainable procurement, sustainable development, circular economy, Industry 4.0, performance measurement, human capital in supply chain and operations; decision modelling for sustainable business, and integration of operation area with others areas.

Jose Arturo Garza-Reyes is a Professor of Operations Management and Head of the Centre for Supply Chain Improvement at the University of Derby, UK. He is actively involved in industrial projects, where he combines his knowledge, expertise, and industrial experience in operations management to help organisations achieve excellence in their internal functions and supply chains. As a leading academic, he has led and managed international research projects funded by the European Commission, British Academy, British Council, and Mexico's National Council of Science and Technology (CONACYT). He has published extensively in leading scientific journals and five books in the areas of operations management and innovation, manufacturing performance measurement, and quality management systems. He is a Co-founder and current Editor of the International Journal of Supply Chain and Operations Resilience (Inderscience), Associate Editor of the International Journal of Production and Operations Management, Associate Editor of the Journal of Manufacturing Technology Management, and Editor-in-Chief of the International Journal of Industrial Engineering and Operations Management. Professor Garza-Reyes has also led and guest edited special issues for Supply Chain Management: An International Journal, International Journal of Lean Six Sigma, International Journal of Lean Enterprise Research, International Journal of Engineering Management and

Economics, and International Journal of Engineering and Technology Innovation. Areas of expertise and interest for Professor Garza-Reyes include general aspects of operations and manufacturing management, business excellence, quality improvement, and performance measurement. He is a Chartered Engineer (CEng), a certified Six Sigma-Green Belt, and has over eight years of industrial experience working as Production Manager, Production Engineer, and Operations Manager for several international and local companies in both the U.K and Mexico. He is also a fellow member of the Higher Education Academy (FHEA) and a member of the Institution of Engineering Technology (MIET).

Syed Abdul Rehman Khan is a Post-Doctoral Researcher in Tsinghua University, China. Dr. Khan achieved his CSCP-Certified Supply Chain Professional certificate from the U.S.A. and successfully completed his PhD in China. He has more than nine years' core experience of supply chain and logistics at industry and academic levels. He has attended several international conferences and also has been invited as keynote speaker in different countries. He has published many scientific research papers in different well-renowned international journals and conferences. In addition, Dr. Khan has achieved scientific innovation awards two times consecutively by the Education Department of Shaanxi Provincial Government, China. Also, Dr. Khan holds memberships in the following well-renowned institutions and supply chain bodies/associations: APCIS-U.S.; Production and Operation Management Society, India; Council of Supply Chain Management of Professionals U.S., Supply Chain Association of Pakistan, and Global Supply Chain Council China.

Contributors

Sreejith Balasubramanian
Middlesex University Dubai
Dubai, UAE

Vikas Kumar Choubey
Department of Mechanical
Engineering
National Institute of Technology
Patna, Bihar, India

Elaine Conway
Derby Business School
University of Derby
Derby, UK

R.A. Dakshina Murthy
Prin. L.N. Welingkar Institute
of Management Development
& Research
Bangalore, Karnataka, India

Aastha Dhoopar
Department of Management
Studies (DMS)
Malaviya National Institute of
Technology (MNIT)
Jaipur, Rajasthan, India

Anukriti Dixit
Public Systems Group
Indian Institute of Management
Ahmedabad, Gujarat, India

Leena N. Fukey
School of Business and Management
CHRIST (Deemed to be University)
Bangalore, Karnataka, India

Mahnoush Gharehdash
Middlesex University Dubai
Dubai, UAE

Mahshad Gharehdash
Middlesex University Dubai
Dubai, UAE

Hadi Lal Ghorbani
Department of Industrial Engineering
Semnan Branch
Islamic Azad University
Semnan, Iran

Ali Jahan
Department of Industrial Engineering
Semnan Branch
Islamic Azad University
Semnan, Iran

Leena James
Deanery of Commerce & Management
Christ Deemed University
Bangalore, Karnataka, India

Mahender Singh Kaswan
School of Mechanical Engineering
Lovely Professional University
Phagwara, Punjab, India

Anil Kumar
Guildhall School of Business and Law
London Metropolitan University
London, UK

Mukesh Kumar
Department of Mechanical
Engineering
National Institute of Technology
Patna, Bihar, India

Fernando León-Mateos
School of Economics and Business
University of Vigo
Vigo, Spain

Lucas López-Manuel
School of Economics and Business
University of Vigo
Vigo, Spain

Rahul S. Mor
Department of Food Engineering
National Institute of Food Technology
    Entrepreneurship and Management
Sonepat, Haryana, India

Simon Peter Nadeem
Centre for Supply Chain Improvement
University of Derby
Derby, UK

Alireza Panahandeh
Department of Industrial Engineering
Semnan Branch
Islamic Azad University
Semnan, Iran

Rajeev Rathi
School of Mechanical Engineering
Lovely Professional University
Phagwara, Punjab, India

Antonio Sartal
School of Economics and Business
University of Vigo
Vigo, Spain

Jeagyung Seong
Gudeul Technology
Wonju, South Korea

Vinaya Shukla
Middlesex University
London, UK

Priyanka Sihag
Department of Management Studies
    (DMS)
Malaviya National Institute of
    Technology (MNIT)
Jaipur, Rajasthan, India

Sarbjit Singh
Department of Industrial & Production
    Engineering
National Institute of Technology
Jalandhar, Punjab, India

Mudita Sinha
School of Business and Management
CHRIST (Deemed to be University)
Bangalore, Karnataka, India

Ashok Kumar Suhag
Department of Electronics and
    Communication
BML Munjal University
Gurgaon, Haryana, India

Nurullah Umarusman
Faculty of Economics & Administrative
    Sciences, Department of Business
    Administration
Aksaray University
Aksaray, Turkey

Arvind Upadhyay
Brighton Business School
University of Brighton
Brighton, UK

Ammar Vakharia
School of Electrical and Electronics
    Engineering
Lovely Professional University
Phagwara, Punjab, India

Shikha Verma
Production and Quantitative Methods
Indian Institute of Management
Ahmedabad, Gujarat, India

Sarika Yadav
Department of FBM & ED
National Institute of Food
    Technology Entrepreneurship
    and Management
Sonepat, Haryana, India

Section 1

Conceptual Understanding and Adoption Challenges of Circular Economy Practices

1 The Conceptual Model Framework for the Role of Human Resources in the Adoption of the Circular Economy
A Content Analysis Approach

Priyanka Sihag and Aastha Dhoopar
Department of Management Studies (DMS),
Malaviya National Institute of Technology
(MNIT), Jaipur, Rajasthan, India

Anil Kumar
Guildhall School of Business and Law, London
Metropolitan University, London, UK

Ashok Kumar Suhag
Department of Electronics and Communication,
BML Munjal University, Gurgaon, Haryana, India

CONTENTS

1.1   INTRODUCTION

The concept of circular economy (CE) is postulated to be a new business outlook, an emerging approach that may assist the organizations and the societies in realizing the goal of sustainable development (McDowall *et al.*, 2017). The production and consumption patterns of the humans across the globe have put the environment into a state of increased risk and precariousness. To address this issue, the adoption of CE proposes a novel perspective in terms of the organizational production and consumption, the perspective that emphasizes on restoration of the value of resources used. It suggests that replacing traditional approach of linear economy ('take, make and dispose') with the circular approach (Jabbour *et al.*, 2018) of energy as well as physical resources can contribute toward economic, environmental and social advantages (Geissdoerfer *et al.*, 2017).

In the linear economy, nearly 80% of what is used is straight away discarded after use (Sempels & Hoffmann, 2013).The waste generated in a linear economy affects human health and the environment, whereas the waste that comes from different processes when inserted into a CE produces 'beneficial artifacts' for human use (Sikdar, 2019), as CE is a production and consumption system that aims at keeping the parts, products, resources and energy in circulation for addition, recreation and maintenance of value over a period of time (Jabbour *et al.*, 2019a).

The discussion around the concept of CE began in the mid-1960s (Murray *et al.*, 2017). The term CE was first introduced by an ecological economist Boulding (1966) and is deeply rooted in the general systems theory (von Bertalanffy, 1950), according to Ghisellini *et al.* (2016). Boulding (1966) in his seminal work has depicted earth as a closed circular system with finite absorbing capacity, and articulated that there should be an equilibrium between the economy and the environment for them to coexist (Geissdoerfer *et al.*, 2017). Since the 1970s, the concept of CE has been gaining momentum (EMF, 2013). Merli *et al.* (2018) have viewed CE as a progressive, multidisciplinary concept that extricates the economic growth from utilization of resources and social implications only. CE is the foundation of a green economy perspective that advances the focus from utilization of materials and associated wastages to welfare of humans and ecosystem resilience (Reichel *et al.*, 2016).

As far as the conception of CE is concerned, there is an absolute lack of consensus. Kirchherr *et al.* (2017) in their extensive work have singled out 114 different definitions of the concept, pinpointing toward the lack of concurrence on the subject. As diversified as the conceptualization of CE might be, the active role of human resources in adoption of CE is undeniable, and this is one of the major gaps in the CE literature that needs to be pursued further. 'Circular Economy', 'Circular Economy and Human Resources', 'HRM and Circular Economy' and 'HR and Sustainability' were used as the key terms to delineate the contribution of human resources toward the adoption of a CE.

The purpose of this study is to ascertain the role that the human resources can play in smooth transition toward the CE in today's times. The prominence of Green HRM (GHRM), Eco-Innovation, Awareness at various levels of the organization and management control have been discussed as a part of the role of human resources. The chapter has been structured as follows: Section 2 discusses the CE – sustainability link, Section 3 incorporates the prominence of adoption of the CE. In Section 4, the role of human resources has been discussed in length followed by conclusion and future research recommendations in Section 5 and 6 respectively.

1.2   THE CE – SUSTAINABILITY LINK

The notion of CE is associated with the sustainability sciences, which is grounded in the research streams of industrial ecology (Erkman, 1997), cleaner production (Fresner, 1998), cradle-to-cradle (C2C) (McDonough & Braungart, 2002), industrial ecosystems (Jelinski *et al.*, 1992), industrial symbioses (Chertow & Ehrenfeld, 2012), biomimicry (Benyus, 1997), regenerative design (Lyle, 1996), performance economy (Stahel, 2010), natural capitalism (Hawken *et al.*, 1999) and the conceit of zero emissions (Pauli, 2010). In the context of business enterprises today, CE is thought-through as a means of better resiliency, cost reduction, creation of value, revenue and legitimacy (Park *et al.*, 2010; Tukker, 2015; Urbinati *et al.*, 2017; Manninen et al., 2018).

As defined by WCED (World Commission on Environment & Development, 1987), Sustainability is the 'development that meets the needs of the present generation without compromising the ability of future generations to meet their ends'. CE is an integral part of sustainable development (Moktadir *et al.*, 2018). It is identified as a closed loop value chain (Preston, 2012), wherein the complete wastage is collected through proper channels and reiterated to the manufacturing units to be reused (Yuan and Moriguichi, 2008). It is that economy whose design happens to be restorative and regenerative, focusing on the expansion of the value chain (Moktadir *et al.*, 2018). It focuses on sustainable manufacturing practices as well as sustainable environmental practices through elimination and reduction of elementary waste (Fischer & Pascucci, 2017).

CE is considered as a part of sustainable development framework, which operates on the principle of 'closing the life cycle' of goods resulting in the minimization of raw materials, energy and water (Jabbour *et al.*, 2019a), and intends to conceive a restorative industrial design (Geissdoerfer *et al.*, 2017). CE is a structure formed by the societal production–consumption systems that magnifies the services produced from the linear nature–society–nature material and energy flow by the use of cyclical material, renewable sources of energy and energy flows of the cascading type (Korhonen *et al.*, 2018). The adoption of CE stimulates people towards more sustainable actions and ensures formulation of regulations that cater to the goal of sustainability (Andersen, 2007; Besio & Pronzini, 2014; Haas *et al.*, 2015; Miliute-Plepiene & Plepys, 2015; Schneider, 2015). CE is the primary driver that gives impetus to a society that is more sustainable (UNEP, 2006), and augments eco-innovations such that the well-being at the social (Geng *et al.*, 2016) and economic level is ensured (Genovese *et al.*, 2017; De Jesus & Mendonça, 2018).

CE and the concept of sustainability are considered similar, but CE is a condition, a beneficial precursor for sustainability (Geissdoerfer *et al.*, 2016). Both the concepts focus on the intra- and inter-generational commitments triggered by ecological hazards, which indicate the significance of deliberation on the concurring pathways of development, both the concepts emphasize on the shared responsibilities and the importance of coordination between multiple players, wherein the system design and the innovativeness are the main drivers for realization of the said goals. While the CE aims at evolution of a closed loop, eliminating wastages and leakages of the system, the concept of sustainability tends to deliver environmental, economic and societal benefits at large (Elkington, 1997), and the ones benefiting from the adoption of CE are the economic players responsible for implementation of the system. The CE is preeminently associated with the economic systems with primary benefits for the environment and tacit benefits for the society; on the other hand, the notion of sustainability, as originally developed, treats all the three dimensions equitably (Geissdoerfer *et al.*, 2016).

1.3 THE ADOPTION OF CIRCULAR ECONOMY

The significance of CE stems from the fact that natural resources are scarce and maximization of the circularity of resources and energy within systems may lead to retention of some value from these resources at the end of their life (Ghisellini *et al.*, 2016). The Natural resources are finite and being depleted ruthlessly, the consumption of resources is more than that can be replaced (Meadows *et al.*, 2004). It is expected that by 2050, the population living on earth and enjoying growing wealth would be around 9 billion (Godfray *et al.*, 2010), which will lead to the demand of resources to almost thrice of what it is currently. CE involves a systems approach where interdependence and holism are of utmost importance to manage the finite resources of the companies (Ünal *et al.*, 2018).

In the World Economic Forum (2014) Report, the Ellen MacArthur Foundation and McKinsey & Company have deduced that the adoption of the CE would generate an opportunity of more than a trillion USD for the worldwide economy. Macroeconomically, the adoption of CE model may result in enhancement of resource productivity by 3% resulting in cost-savings of 0.6 trillion euros per year, addition of 1.8 trillion euros to other economic benefits by 2030 (McKinsey & Company, 2015) and net material cost and saving benefits of more than 600 billion USD p.a. by 2025 through this restorative approach (The Ellen MacArthur Foundation, 2013). The World Business Council for Sustainable Development (2017) has asserted that the adoption of the CE can result in increased growth, competitive advantages and innovation along with reduction in costs, energy use and emission leading to a better supply chain and judicious use of resources. Even PricewaterhouseCoopers (2017a) has adjudged that 'the CE is here to stay' and the company has elucidated that the CE will open up the avenues to build competitive advantage and create profit pools, develop resilience and address significant issues faced by the businesses today (PricewaterhouseCoopers, 2017b).

In the literature, three levels of initiatives have been identified for the implementation of the principles of CE: the micro level of firms that relates to the initiatives specific to the firms classified on the basis of 3R's – reduce, reuse and recycle (Ying & Li-Jun, 2012), the second level is the meso level that incorporates the execution of eco-industrial parks, networks and inter-firm collaborations for optimum utilization of resources. Finally, at the macro-level, the initiatives undertaken by the government and policy makers are accounted for (Geng & Doberstein, 2008). Today, the countries are becoming self-reliant and aware of the requirement to switch to a newer system based on the principles of circularity (Bonviu, 2014).

According to MacArthur *et al.* (2015), the three principles that guide the CE cycles are as follows:

- Strengthening the circularity of resources and energy by increasing the life of the resources either through biological or technical cycles.
- Decrement in the negative effects of production setup.
- Conservation of natural resources, i.e. equilibrium between consumption of renewables and the non-renewables.

For implementation of the principles of CE, Ellen MacArthur Foundation (2015) has delineated six business actions – the ReSOLVE Framework:

- *Regenerate* – Based on the adoption of renewable resources and energy, reclaim, retain ecosystems wellness and enhance the natural bio capacity.
- *Share* – According to the shared economy perspective, the significance of ownership is lost when the resources are shared between individuals. Resultantly, products should be designed such that they can be reused and they last longer.
- *Optimize* – It is a technology centered approach. This strategy recommends use of digital manufacturing technologies, like sensors, RFID, big data and remote route to scale down the waste generated in production systems across supply chains in an organization.
- *Loop* – This stems from the biological and technical cycles wherein the components and materials are kept in closed loops and emphasized on the inner loops.
- *Virtualize* – The focus of this strategy is on the replacement of physical products with virtual products.
- *Exchange* – Adoption of new technologies to improve the way in which the goods and services are produced. This implicates replacing obsolete and non-renewable goods with the newer renewable goods.

Apart from the ReSOLVE framework, the principles of CE can be implemented through the 3R's – Reduce, Reuse, Recycle (Wu *et al.*, 2014) or the 6R's – Reuse, Recycle, Redesign, Remanufacture, Reduce, Recover (Jawahir & Bradley, 2016) (Figure 1.1). The CE concept has altered the policies and innovation in the major

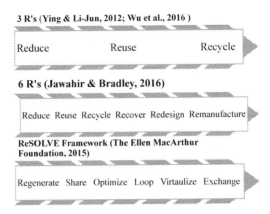

FIGURE 1.1 The different frameworks found in the literature to effectuate the principles of circular economy.

economies of the world, namely China, Germany, Japan and the UK. The extant literature on the circular economies and instances of the successes and breakdowns from the real world necessitate the integration of bottom-up and top-down strategies for implementation and evaluation of CE (Winans *et al.*, 2017).

1.4 THE ROLE OF HUMAN RESOURCES

The 'human side' of the organizations, also referred as the 'soft side' (Wilkinson, 1992), has been of great interest to scholars since a long time (McGregor, 1966). The sustainable human resources (Huselid, 1995) and the CE both exert an influence on the firms' performance (Despeisse *et al.*, 2017) and the competitive advantages, but unfortunately the human side of the CE has not received adequate attention in the CE literature (Jabbour *et al.*, 2019b). The two relevant organizational theories that assist in building sustainable organizations is the stakeholder theory and the resource-based view (RBV) (Sodhi, 2015). The stakeholder theory (Freeman *et al.*, 2004) features HRM as a key factor in influencing and getting influenced by the organizational sustainability management (Sodhi, 2015). Similarly, the RBV of the firm (which identifies a human resource as valuable and unique) (Barney, 2001) and the natural RBV uphold the positive effects of HRM (Wright *et al.*, 1994) and sustainability initiatives (Hart and Dowell, 2011) on firm performance respectively. The human side or the 'soft side' of the organizations is as important as the hard or technical side when the question of managing an organization for organizational and environmental sustainability comes forth (Renwick *et al.*, 2013).

In the purview of human resource management, employees are the critical factors that are instrumental to value addition in an organization, and these are the resources that are capable of making a difference when it comes to innovation, organizational performance and the eventual business success (Bakker and Schaufeli, 2008). Human capital is the most important asset for any organization (Guest, 2001), because it is owing to the capability of human resources that the physical resources are utilized and the income is generated for the business firms (Flamholtz, 1999).

In order to amalgamate the concepts of human resource management and CE (Figure 1.2), the formulation of such strategies is of utmost importance that stimulate the economic, social and environmental strategies of the organizations (Jabbour & Santos, 2008).

The contribution of human resource management towards CE and sustainability can be seen in the following ways: Innovation helps the organizations find superior

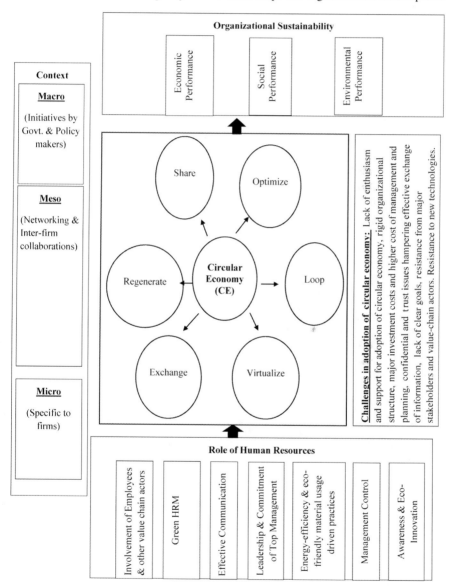

FIGURE 1.2   The conceptual model depicting the role of human resources in adoption of the circular economy along with the challenges and CE as a precursor to organizational sustainability are also outlined.

performance in economic terms (Jamrog *et al.*, 2006), the effective management of diversity can contribute towards social performance and effectiveness of the environmental management system and the augmentation of eco-friendly products can contribute towards environmental performance (Jabbour and Santos, 2008). The ways in which human resources can play a role in adoption of CE is detailed as follows.

1.4.1 Green HRM

GHRM is critical in establishing desirable sustainability initiatives for organizations (Jackson *et al.*, 2014). GHRM practices encompass various dimensions like eco-focused recruitment and selection, green performance measurement and rewards, ecological training, developing green teams in the organization, empowerment of employees in relation to environmental initiatives, and cultivating an organizational culture with major focus on environmental sustainability (Jabbour & Santos, 2008). The effect of GHRM on firms' sustainability indicators is an outcome of the workforce's enhanced green behavior owing to GHRM (Pham *et al.*, 2019; Kim *et al.*, 2019). Green human resource management pertains to the alignment of human resource practices, systems and the strategic aspects like organizational culture and employee empowerment with organizational and environmental goals (Renwick *et al.*, 2013).

The organizations should channel their recruitments towards selection of the people committed to the environmental system, training and evaluation of the performance of the individuals should be done on the basis of environmental criteria, the reward system of the company as well should incorporate remunerative and non-remunerative ways to stimulate environmental performance, inculcate environmental values as a part of corporate culture and promote environmental education and interaction between the teams to combat the environmental problems (Wehrmeyer, 1996; Jabbour and Santos, 2008).

The accomplishment of the goal of CE is not the responsibility of the human resources at a particular level, the employees as well as the management irrespective of the organizational hierarchy should aim towards the realization of CE. Similarly, the introduction of sustainability measures should be accompanied by attempts to embed them within the internal–external relationships framework, the reward systems and more, if not so the initiatives are bound to be fruitless. On similar grounds, the introduction to CE needs to be accompanied by the concept of 'Green HRM' (Kirsch and Connell, 2018).

1.4.2 Effective Communication and Active Involvement of Value-Chain Actors

Over the past years, the insertion of the environmental dimension in the everyday organizations is the biggest change that has taken place in the business world (Rosen, 2001). The model of CE calls for innovation (Singh and Ordoñez, 2016) and new skills. For an economy to adopt the circular design, it is imperative for it to develop new proficiencies, along with the evolution in the skill-set for smooth transition towards the CE, increased attention should be given to the participation of all value-chain actors (Bocken *et al.*, 2018) and effective communication across the supply chain, which can be done by enhancing trust and shared values collectively to attain the intended goal (Ünal *et al.*, 2018). The dissemination of environment related information in an organization should appease the benchmarks of timeliness, relevance, accessibility and precision (Soo Wee and Quazi, 2005).

Stahel (2016) brought in the notion of performance economy that forms the basis for business models of CE. Instead of sale of the product, the sale of use of the product is gaining popularity. The organizations should endorse the concept of CE in their marketing activities as well (Kumar & Venkatesan, 2005; Van Heerde *et al.*, 2013; Baxendale *et al.*, 2015), which can be done through the advertisements on the company website, in-store advertising and sales personnel, communication around CE through various channels and association of customers with the circularity initiatives of the organization.

1.4.3 LEADERSHIP AND COMMITMENT FROM THE TOP MANAGEMENT

Commitment can be described as a psychological state that delineates the interaction of individuals with various aspects of the company, affecting the final decision to keep up the connection with the aspect in consideration (Lämsä and Savolainen, 2000; Meyer *et al.*, 2002). Managerial commitment is strategically important for the alignment of resources with the pre-determined objectives of the company (Ünal *et al.*, 2018). Apart from the employees of the organizations, it is the top leadership that has the responsibility to maintain a sustainable environment. The top management may play an important part in adopting cleaner technologies for the process of manufacturing in companies (Ghazilla *et al.*, 2015), accommodation of such techniques in the manufacturing process leads to reduced levels of water pollution, less carbon emissions, eco-friendly supply systems, green transportation and sustainable manufacturing practices in a CE (Nowosielski *et al.*, 2007). The setting of environmental vision and policies is the most important responsibility of the top management in today's times. To attain the environmental goals, be it the goal of sustainability or the adoption of CE, the environmental issues should be integrated into the significant business functions and operations (Soo Wee & Quazi, 2005).

At the managerial level, planning, organizing, leading and controlling are the major functions of management, for the adoption of CE, effective planning and management for optimum utilization of resources is required, inappropriate planning or the lack of preparation may largely hamper the adequacy of CE and mislead the active players across the supply chain (Mangla *et al.*, 2018).

As evident in the literature, there is a lack of managerial support to environmental practices (Zhu & Geng, 2013), generally it is seen that the environmental practices are actualized in a top-down manner with a major responsibility to attain the goal of sustainability restricted to the CEOs as these are the human resources who have the utmost influence on the allocation of resources and strategy formulation (Kiron *et al.*, 2012; Epstein & Buhovac, 2014).

1.4.4 ECO-INNOVATION

The process of innovation involves increase in the value of the organizations through the development of either new knowledge or processes for the effective and dynamic use of knowledge available for use or to facilitate organizational change. Furthermore, human resource management has a crucial role to play in encouraging innovation inside the organizations (Jabbour & Santos, 2008). The positive impact of HRM on innovative performance of the companies has been empirically supported by the study of Laursen and Foss (2003).

As pointed out by Schuler and Jackson (1987), the major aim of HRM is to foster innovation and put in action the practices that compel the employees to think and create in different ways. Long-term focus, higher risk-taking abilities, higher degree of cooperation and interdependent behavior are a few characteristics suggested by the above-mentioned authors for the alignment of employees' behavior with the innovation strategy of the company.

The advancements of cleaner technologies especially require human resources to enact a diligent role to be aware of the environmental strategy of the company, and the competencies required for the enhancement of the environmental performance. For the environmental dimension to be included in the fundamental organizational activities, employees are the primary players engaged in this task (Rothenberg, 2003).

1.4.5 MANAGEMENT CONTROL

Management control, as introduced by Robert Anthony, establishes a link between the strategic planning and the operational control (Herath, 2007). The notable ambit of the management control lies in (a) evolution, analysis and response to the information for critical decision-making and (b) to direct the employees' behavior such that the way in which the employees behave and the decisions are coherent with the overall organizational objectives and strategies (Chenhall, 2003; Anthony *et al.*, 2014). The use of management control is vital for the organization to attain its strategic objective of circularity (Svensson & Funck, 2019). For adapting to CE, Malmi and Brown's (2008) framework enumerating five management control mechanisms has been discussed below:

- *Culture controls* includes articulation of organizational values and visions, socialization of employees post recruitment to guide and acquaint them with the organizational procedures and existing manpower. The company's vision regarding the circular values should be clearly communicated. Internal communication of circularity is essential for the employees, but external cultural control for disseminating the idea of CE to foster the demand of sustainable products and services is even more important (Uyarra *et al.*, 2014).
- *Planning* signifies planning at the strategic level and action planning. The strategies for adopting CE should be translated into specific action plans, but this action planning is not possible without effective internal communication in the organization.

- *Cybernetic controls* constitute the budgeting, measurement and cost accounting systems. When planning a product, the organizations should emphasize on the flexibility of its use and product life extension, which provides cost benefits and affects investment appraisals (Svensson & Funck, 2019).
- *Rewards and Compensation* to keep the employees motivated and to make them feel valued in the organization.
- *Administrative Controls* such as division of work and delegation of authority and responsibility to make decisions. All these mechanisms are related to each other as concrete action plans, analysis of costs and material flows, all these functions affect the ultimate decision making in an organization.

1.4.6 AWARENESS ABOUT THE CIRCULAR ECONOMY

Employee involvement and motivation in the organization are important factors in gaining knowledge about the CE. The employees may come forward and apprise the management about the benefits of CE (Moktadir *et al.*, 2018). Taking the perspective of customers into consideration, in today's times the consumers are more concerned about the environmental outcomes. Awareness of the customers about the green initiatives plays a pivotal role in the adoption of CE (Stock and Seliger, 2011). The execution of 'Green' teams is a critical step towards the realization of environmental goals. The awareness possessed by the employees is not enough, the awareness and the knowledge of the CE and the environmental dimensions possessed by them should be utilized within the organizations. The employees should be motivated to give suggestions, the inputs offered by them should be valued and the employees should be empowered to handle the circumstantial problems and contribute towards the environmental performance of the company (Soo Wee and Quazi, 2005).

1.4.7 ENERGY EFFICIENCY AND ECO-FRIENDLY MATERIAL USAGE-DRIVEN PRACTICES

There have been many attempts by the scholars in the past to assess indicators of energy efficiency and conservation in reference to a CE (Li *et al.*, 2010; Su *et al.*, 2013). The main focus of a CE is to minimize the amount of energy consumed and to reduce leakages in the system (Stahel, 2013). The production processes and products should be designed in such a way that the detrimental impacts on the environment are minimized. The environmental concerns should be integrated into the product at the design stage only and the recycling activities should be planned too to warrant full usage of resources (Soo Wee & Quazi, 2005).

The management and the employees of the enterprises engaged in the process of manufacturing should strive to make the business plan feasible by using ad-hoc strategies (natural, dependable, reusable and divisible) of handling the material in the initial stages of product development itself (Ünal *et al.*, 2018). In addition to this, the role of design (DfX practices) (design for reuse, design for remanufacturing, design for environment, etc.) has been accounted as an important factor for the transition towards an economy that is more circular (Moreno *et al.*, 2016; De los Rios & Charnley, 2017).

CE is the exemplification of the quintessential shift that considers nature as an inspiration to respond to social and environmental needs; it is a system that warrants a paradigm switch in the way the goods and services are produced and consumed (Cohen-Rosenthal, 2000; Hofstra & Huisingh, 2014). The adoption of CE can undoubtedly contribute in making the processes as well as the products more efficient. The human resources especially the top management has a pivotal role to play in the adoption of the CE, the human resources should emphasize on practicing a more closed-loop recycling where the focus is on the reusage of material (the materials like glass and steel can be recycled constantly). In addition to embracing the closed-loop recycling, the use of products should be widened. The human resources are endowed with a special feature that is knowledge, it is through the application of this knowledge that the organizational goals and development can be realized. The human resources should convene such policies, which aim at lengthening and widening the usage of products, wherein the remanufacturing of physical goods takes place such that the goods can be reutilized for secondary markets or the industries running with less sophisticated infrastructure. Along with taking up the remanufacturing practices, the organizations instead of selling products to the customers directly, renting or leasing may be opted for, under which the ownership of the goods is retained by the company throughout the life cycle of the product and upon exhaustion of the product, it may well be returned to the organization itself (Tse *et al.*, 2016).

1.5   CONCLUSION

Ever since the industrial revolution, the human race has been excessively reliant on the natural resources for consumption and improvisation in the standards of living, but with elapse of time the resources are becoming increasingly scarce and expensive; to deal with this problem of limited resources, new ways to conceive a more sustainable environment should be brought into being rapidly (Tse *et al.*, 2016). CE has a resolute environmental focus (Jones & Comfort, 2017) and is a pertinent strategy that contemplates a new way to modify the traditional system focusing only on the consumption at customers end into a circular system (Stahel, 2013). CE is envisioned as an approach to lessen the conflicts between competitive and environmental preferences of a company, shaping the organization to be more competitive and reducing its environmental impressions at the same time (Gusmerotti *et al.*, 2019).

CE is viewed as a promising solution for a number of reasons. Primarily, the adoption of CE minimizes the overexploitation of natural resources and waste generation. Secondly, across the life cycles of products, the control over the goods and materials lies with the focal firms only whereby the firms can choose to retain the ownership of the product with themselves and offer it as a service to the customers (Bocken *et al.*, 2016).The major focus of the corporate initiatives and the human resources while framing the policies has been on recycling and zero waste procedures (Jones & Comfort, 2017), but the development of an effective CE postulates the inclusion of whole consumer product life cycle and waste management so as to draw in all the sectors of the economy into the realm of circularity.

The literature around the CE constitutes of the conceptualization of the CE, determinants of CE, principles to guide adoption of CE and barriers to the implementation of circular economy. The significance of organizations adopting a CE is very clear in the literature but the way in which this can be actually done is obscure. Since, the transition to the CE requires involvement of human resources at various levels and organizational functions, the role of human resource management in supporting the CE is manifold. Even Eisenstat (1996) has asserted that human resources have a dominant role to play in organizations as it is a function that can vitalize the issues of sustainability by incorporating those in the inter-firm as well as intra-firm relationships.

According to Geissdoerfer *et al.* (2017), CE can be achieved through maintenance, remanufacturing, recycling, facilitating long-lasting arrangement, repair, reuse and refurbishing, but the execution of all these measures is not possible without the intervention of human resources. The effective implementation of the CE is only possible through the amalgamation of advanced technologies into the processes (Su *et al.*, 2013), the choice of technology involved also points towards the importance of human resources in the adoption of CE. As mechanized and automated as the processes in today's world might be, the effectiveness and efficiency of the organizational activities and procedures depends on the adequacy and alignment of the natural, technical and human resources.

Adoption of CE at the organizational level is undoubtedly a complicated task (Svensson and Funck, 2019) and the transition to CE in organizations requires novel strategies to conduct business activities (ING, 2015), requiring new targets to be calibrated in new ways (EASAC, 2016). It is of primal importance for the human resources of an organization to support the environmental management system of the organizations, foster organizational change and ensure alignment of all the functional dimensions (Wehrmeyer, 1996).

1.6 FUTURE RESEARCH DIRECTIONS

Since the literature on the CE is extensive, there are a number of research directions that need to be addressed for effective implementation of the CE. Exchange of information has been one of the major constraints on the efficacy of CE (Winans *et al.*, 2017), future research should incorporate the communication and involvement within the organizations as the central point of research to cater to the requirements of CE. Secondly, the integrative framework assimilating the role of human resources in this chapter may be empirically tested to know how the variables interact with each other as there is an absolute lack of empirical studies on the link between human resources and CE, further research on the subject can be substantiated through case studies. Thirdly, the CE connotes radical changes in the managerial practices of organizations; for example, utilizing resources and energy in an efficient way such that the environmental impact is reduced, more attention should be given to the ways in which the companies can integrate the scarce resources with the CE managerial capabilities (Ünal *et al.*, 2018). On similar grounds, which all managerial practices should be adopted by the companies for the implementation of the CE warrants more attention. Moreover, research on CE should focus on initiatives taken by people, the contribution of these towards the triple bottom line (Elkington, 1997). Attention should also be given to organizational sustainability in this context. In the adoption of CE, multiple players are involved that too at various levels, integration of the actions of all the players is challenging and elimination of resistance on the part of stakeholders, customers and top management may be difficult. The competence of human resources in promoting and embracing the sustainability initiatives such as the CE span across the boundaries of a firm (Nejati *et al.*, 2017), and given the rapidly changing environment that the humans are living in, environmental sustainability (Jackson *et al.*, 2014), environmental management and the CE should be the focal point of the budding HR research.

REFERENCES

Andersen, M. S. (2007). An introductory note on the environmental economics of the circular economy. *Sustainability Science*, 2(1), 133–140.

Anthony, R. N., Govindarajan, V., Hartmann, F. G., Kraus, K., & Nilsson, G. (2014). *Management control systems: European edition*. UK: McGraw-Hill Education.

Bakker, A. B., & Schaufeli, W. B. (2008). Positive organizational behavior: Engaged employees in flourishing organizations. *Journal of Organizational Behavior: The International Journal of Industrial, Occupational and Organizational Psychology and Behavior*, 29(2), 147–154.

Barney, J. B. (2001). Is the resource-based "view" a useful perspective for strategic management research? Yes. *Academy of Management Review*, 26(1), 41–56.

Baxendale, S., Macdonald, E. K., & Wilson, H. N. (2015). The impact of different touchpoints on brand consideration. *Journal of Retailing*, 91(2), 235–253.

Benyus, J. M. (1997). *Biomimicry: Innovation inspired by nature*. New York: William Morrow and Company.

Besio, C., & Pronzini, A. (2014). Morality, ethics, and values outside and inside organizations: An example of the discourse on climate change. *Journal of Business Ethics*, 119(3), 287–300.

Bocken, N. M., De Pauw, I., Bakker, C., & van der Grinten, B. (2016). Product design and business model strategies for a circular economy. *Journal of Industrial and Production Engineering*, 33(5), 308–320.

Bocken, N. M., Schuit, C. S., & Kraaijenhagen, C. (2018). Experimenting with a circular business model: Lessons from eight cases. *Environmental Innovation and Societal Transitions*, 28, 79–95.

Bonviu, F. (2014). The European economy: From a linear to a circular economy. *Romanian Journal of European Affairs*, 14, 78.

Boulding, K. E. (1966). *The economics of the coming spaceship earth*. In H. Jarrett (ed.) Environmental Quality in a Growing Economy. Baltimore, MD: Resources for the Future/John Hopkins University Press (pp. 3–14).

Chenhall, R. H. (2003). Management control systems design within its organizational context: Findings from contingency-based research and directions for the future. *Accounting, Organizations and Society*, 28(2–3), 127–168.

Chertow, M., & Ehrenfeld, J. (2012). Organizing self-organizing systems: Toward a theory of industrial symbiosis. *Journal of Industrial Ecology*, 16(1), 13–27.

Cohen-Rosenthal, E. (2000). A walk on the human side of industrial ecology. *American Behavioral Scientist*, 44(2), 245–264.

De Jesus, A., & Mendonça, S. (2018). Lost in transition? Drivers and barriers in the eco-innovation road to the circular economy. *Ecological Economics*, 145, 75–89.

De los Rios, I. C., & Charnley, F. J. (2017). Skills and capabilities for a sustainable and circular economy: The changing role of design. *Journal of Cleaner Production, 160*, 109–122.

Despeisse, M., Baumers, M., Brown, P., Charnley, F., Ford, S. J., Garmulewicz, A., Knowles, S., Minshall, T.H.W., Mortara, L., Reed-Tsochas, F. P., & Rowley, J. (2017). Unlocking value for a circular economy through 3D printing: A research agenda. *Technological Forecasting and Social Change, 115*, 75–84.

EASAC (2016). Indicators for a Circular Economy. EASAC Policy Report 30. November. EASAC. Available at https://easac.eu/fileadmin/PDF_s/reports_statements/Circular_Economy/EASAC_Indicators_web_complete.pdf (Accessed *17 May, 2020*).

Eisenstat, R. A. (1996). What corporate human resources brings to the picnic: Four models for functional management. *Organizational Dynamics, 25*(2), 7–22.

Elkington, J. (1997). *Cannibals with Forks: The triple bottom line of 21st century business.* Capstone: Oxford.

Epstein, M. J., & Buhovac, A. R. (2014). *Making sustainability work: Best practices in managing and measuring corporate social, environmental, and economic impacts.* San Francisco: Berrett-Koehler Publishers.

Erkman, S. (1997). Industrial ecology: An historical view. *Journal of Cleaner Production, 5*(1–2), 1–10.

Fischer, A., & Pascucci, S. (2017). Institutional incentives in circular economy transition: The case of material use in the Dutch textile industry. *Journal of Cleaner Production, 155*, 17–32.

Flamholtz, E. (1999). *Human resource accounting: Advances in concepts, methods, and applications.* New York, NY: Springer.

Freeman, R. E., Wicks, A. C., & Parmar, B. (2004). Stakeholder theory and "the corporate objective revisited". *Organization Science, 15*(3), 364–369.

Fresner, J. (1998). Cleaner production as a means for effective environmental management. *Journal of Cleaner Production, 6*(3–4), 171–179.

Geissdoerfer, M., Bocken, N. M., & Hultink, E. J. (2016). Design thinking to enhance the sustainable business modelling process – A workshop based on a value mapping process. *Journal of Cleaner Production, 135*, 1218–1232.

Geissdoerfer, M., Savaget, P., Bocken, N. M., & Hultink, E. J. (2017). The circular economy: A new sustainability paradigm? *Journal of Cleaner Production, 143,* 757–768.

Geng, Y., & Doberstein, B. (2008). Developing the circular economy in China: Challenges and opportunities for achieving 'leapfrog development'. *The International Journal of Sustainable Development & World Ecology, 15*(3), 231–239.

Geng, Y., Sarkis, J., & Ulgiati, S. (2016). Sustainability, well-being, and the circular economy in China and worldwide. *Science, 6278*, S73–S76.

Genovese, A., Acquaye, A. A., Figueroa, A., & Koh, S. L. (2017). Sustainable supply chain management and the transition towards a circular economy: Evidence and some applications. *Omega, 66*, 344–357.

Ghazilla, R. A. R., Sakundarini, N., Abdul-Rashid, S. H., Ayub, N. S., Olugu, E. U., & Musa, S. N. (2015). Drivers and barriers analysis for green manufacturing practices in Malaysian SMEs: A preliminary findings. *Procedia Cirp, 26*(1), 658–663.

Ghisellini, P., Cialani, C., & Ulgiati, S. (2016). A review on circular economy: The expected transition to a balanced interplay of environmental and economic systems. *Journal of Cleaner Production, 114*, 11–32. https://doi.org/10.1016/j.jclepro.2015.09.007

Godfray, H. C. J., Beddington, J. R., Crute, I. R., Haddad, L., Lawrence, D., Muir, J. F., Pretty, J., Robinson, S., Thomas, S. M., & Toulmin, C. (2010). Food security: The challenge of feeding 9 billion people. *Science, 327*(5967), 812–818.

Guest, D. (2001). *Voices from the boardroom.* London: Chartered Institute of Personnel and Development.

Gusmerotti, N. M., Testa, F., Corsini, F., Pretner, G., & Iraldo, F. (2019). Drivers and approaches to the circular economy in manufacturing firms. *Journal of Cleaner Production, 230,* 314–327. https://doi.org/10.1016/j.jclepro.2019.05.044

Haas, W., Krausmann, F., Wiedenhofer, D., & Heinz, M. (2015). How circular is the global economy?: An assessment of material flows, waste production, and recycling in the European Union and the world in 2005. *Journal of Industrial Ecology, 19*(5), 765–777.

Hart, S. L., & Dowell, G. (2011). Invited editorial: A natural-resource-based view of the firm: Fifteen years after. *Journal of Management, 37*(5), 1464–1479.

Hawken, P., Lovins, A., & Lovins, H. L. (1999). Natural Capitalism: Creating the Next Industrial Revolution. New York: Little, Brown & Company.

Herath, S. K. (2007). A framework for management control research. *Journal of Management Development, 26*(9), 895–915.

Hofstra, N., & Huisingh, D. (2014). Eco-innovations characterized: A taxonomic classification of relationships between humans and nature. *Journal of Cleaner Production, 66,* 459–468.

Huselid, M. A. (1995). The impact of human resource management practices on turnover, productivity, and corporate financial performance. *Academy of Management Journal, 38*(3), 635–672.

ING (2015). Rethinking Finance in a circular Economy. Available at https://www.ing.nl/media/ING_EZB_Financing-the-Circular-Economy_tcm162-84762.pdf (Accessed *17 May, 2020*).

Jabbour, Charbel José Chiappetta, & Santos, F. C. A. (2008). The central role of human resource management in the search for sustainable organizations. *The International Journal of Human Resource Management, 19*(12), 2133–2154. https://doi.org/10.1080/09585190802479389

Jackson, S. E., Schuler, R. S., & Jiang, K. (2014). An aspirational framework for strategic human resource management. *The Academy of Management Annals, 8*(1), 1–56.

Jackson, S. E., Schuler, R. S., & Jiang, K. (2014). An aspirational framework for strategic human resource management. *The Academy of Management Annals, 8*(1), 1–56.

Jamrog, J., Vickers, M., & Bear, D. (2006). Building and sustaining a culture that supports innovation. *People and Strategy, 29*(3), 9–19.

Jawahir, I. S., & Bradley, R. (2016). Technological elements of circular economy and the principles of 6R-based closed-loop material flow in sustainable manufacturing. *Procedia Cirp, 40*(1), 103–108.

Jelinski, L. W., Graedel, T. E., Laudise, R. A., McCall, D. W., & Patel, C. K. (1992). Industrial ecology: Concepts and approaches. *Proceedings of the National Academy of Sciences, 89*(3), 793–797.

Jones, P., & Comfort, D. (2017). Towards the circular economy: A commentary on corporate approaches and challenges. *Journal of Public Affairs, 17*(4), e1680. https://doi.org/10.1002/pa.1680

Kim, Y. J., Kim, W. G., Choi, H. M., & Phetvaroon, K. (2019). The effect of green human resource management on hotel employees' eco-friendly behavior and environmental performance. *International Journal of Hospitality Management, 76,* 83–93.

Kirchherr, J., Reike, D., & Hekkert, M. (2017). Conceptualizing the circular economy: An analysis of 114 definitions. *Resources, Conservation and Recycling, 127,* 221–232. https://doi.org/10.1016/j.resconrec.2017.09.005

Kiron, D., Kruschwitz, N., Haanaes, K., & von Streng Velken, I. (2012). Sustainability nears a tipping point. *MIT Sloan Management Review, 53*(2), 69–74.

Kirsch, C., & Connell, J. (2018). Organizational Sustainability—Why the Need for Green HRM? In J. Connell, R. Agarwal, Sushil, & S. Dhir (Eds.), *Global value chains, flexibility and sustainability* (pp. 223–239). Singapore: Springer. https://doi.org/10.1007/978-981-10-8929-9_15

Korhonen, J., Honkasalo, A., & Seppälä, J. (2018). Circular economy: The concept and its limitations. *Ecological Economics, 143*, 37–46. https://doi.org/10.1016/j.ecolecon.2017.06.041

Kumar, V., & Venkatesan, R. (2005). Who are the multichannel shoppers and how do they perform?: Correlates of multichannel shopping behavior. *Journal of Interactive Marketing, 19*(2), 44–62.

Lämsä, A. M., & Savolainen, T. (2000). The nature of managerial commitment to strategic change. *Leadership & Organization Development Journal, 21*(6), 297–306.

Laursen, K., & Foss, N. J. (2003). New human resource management practices, complementarities and the impact on innovation performance. *Cambridge Journal of Economics, 27*(2), 243–263.

Li, H., Bao, W., Xiu, C., Zhang, Y., & Xu, H. (2010). Energy conservation and circular economy in China's process industries. *Energy, 35*(11), 4273–4281.

Jabbour, A. B. Lopes de Sousa, Jabbour, C. J. C., Godinho Filho, M., & Roubaud, D. (2018). Industry 4.0 and the circular economy: A proposed research agenda and original roadmap for sustainable operations. *Annals of Operations Research, 270*(1–2), 273–286. https://doi.org/10.1007/s10479-018-2772-8

Jabbour, A. B. Lopes de Sousa, Rojas Luiz, J. V., Rojas Luiz, O., Jabbour, C. J. C., Ndubisi, N. O., Caldeira de Oliveira, J. H., & Junior, F. H. (2019a). Circular economy business models and operations management. *Journal of Cleaner Production, 235*, 1525–1539. https://doi.org/10.1016/j.jclepro.2019.06.349

Jabbour, C. J. Chiappetta, Sarkis, J., Lopes de Sousa Jabbour, A. B., Scott Renwick, D. W., Singh, S. K., Grebinevych, O., Kruglianskas, I., & Filho, M. G. (2019b). Who is in charge? A review and a research agenda on the 'human side' of the circular economy. *Journal of Cleaner Production, 222*, 793–801. https://doi.org/10.1016/j.jclepro.2019.03.038

Lyle, J. T. (1996). *Regenerative design for sustainable development.* New York, NY: John Wiley & Sons.

MacArthur, E., Zumwinkel, K., & Martin, R. S. (2015). *Growth within. A circular economy vision for a competitive Europe.* Ellen MacArthur Foundation.

Malmi, T., & Brown, D. A. (2008). Management control systems as a package—Opportunities, challenges and research directions. *Management Accounting Research, 19*(4), 287–300.

Mangla, S. K., Luthra, S., Mishra, N., Singh, A., Rana, N. P., Dora, M., & Dwivedi, Y. (2018). Barriers to effective circular supply chain management in a developing country context. *Production Planning & Control, 29*(6), 551–569. https://doi.org/10.1080/09537287.2018.1449265

Manninen, K., Koskela, S., Antikainen, R., Bocken, N., Dahlbo, H., & Aminoff, A. (2018). Do circular economy business models capture intended environmental value propositions? *Journal of Cleaner Production, 171*, 413–422.

McDonough, W., & Braungart, M. (2002). *Remaking the way we make things: Cradle to cradle.* New York: North Point Press. *ISBN, 1224942886*, 104.

McDowall, W., Geng, Y., Huang, B., Barteková, E., Bleischwitz, R., Türkeli, S., & Doménech, T. (2017). Circular economy policies in China and Europe. *Journal of Industrial Ecology, 21*(3), 651–661.

McGregor, D. (1966). *The human side of enterprise.* In Leavitt H.J., Pondy L.R., and Boje D.M. (Eds.) Readings in Managerial Psychology (3rd Edition), Chicago and London: The university of Chicago Press.

McKinsey & Company (2015). Europe's circular – economy opportunity. Available at https://www.mckinsey.com/~/media/McKinsey/Business%20Functions/Sustainability/Our%20Insights/Europes%20circular%20economy%20opportunity/Europes%20circulareconomy%20opportunity.ashx (Accessed *18 May, 2020*).

Meadows, D. H., Randers, J., & Meadows, D. L. (2004). *Limits to growth: 30-year update.* New York, NY: Earthscan.

Merli, R., Preziosi, M., & Acampora, A. (2018). How do scholars approach the circular economy? A systematic literature review. *Journal of Cleaner Production, 178*, 703–722.

Meyer, J. P., Stanley, D. J., Herscovitch, L., & Topolnytsky, L. (2002). Affective, continuance, and normative commitment to the organization: A meta-analysis of antecedents, correlates, and consequences. *Journal of Vocational Behavior, 61*(1), 20–52.

Miliute-Plepiene, J., & Plepys, A. (2015). Does food sorting prevents and improves sorting of household waste? A case in Sweden. *Journal of Cleaner Production, 101*, 182–192.

Moktadir, M. A., Rahman, T., Rahman, M. H., Ali, S. M., & Paul, S. K. (2018). Drivers to sustainable manufacturing practices and circular economy: A perspective of leather industries in Bangladesh. *Journal of Cleaner Production, 174*, 1366–1380. https://doi.org/10.1016/j.jclepro.2017.11.063

Moreno, M., De los Rios, C., Rowe, Z., & Charnley, F. (2016). A conceptual framework for circular design. *Sustainability, 8*(9), 937.

Murray, A., Skene, K., & Haynes, K. (2017). The circular economy: An interdisciplinary exploration of the concept and application in a global context. *Journal of Business Ethics, 140*(3), 369–380.

Nejati, M., Rabiei, S., & Jabbour, C. J. C. (2017). Envisioning the invisible: Understanding the synergy between green human resource management and green supply chain management in manufacturing firms in Iran in light of the moderating effect of employees' resistance to change. *Journal of Cleaner Production, 168*, 163–172.

Nowosielski, R., Babilas, R., & Pilarczyk, W. (2007). Sustainable technology as a basis of cleaner production. *Journal of Achievements in Materials and Manufacturing Engineering, 20*(1–2), 527–530.

Park, J., Sarkis, J., & Wu, Z. (2010). Creating integrated business and environmental value within the context of China's circular economy and ecological modernization. *Journal of Cleaner Production, 18*(15), 1494–1501.

Pauli, G. A. (2010). *The blue economy: 10 years, 100 innovations, 100 million jobs.* New Mexico: Paradigm Publications.

Pham, N. T., Tučková, Z., & Jabbour, C. J. C. (2019). Greening the hospitality industry: How do green human resource management practices influence organizational citizenship behavior in hotels? A mixed-methods study. *Tourism Management, 72*, 386–399.

Preston, F. (2012). *A global redesign?: Shaping the circular economy.* London: Chatham House.

PricewaterhouseCoopers (2017a). Spinning around: Taking control in a circular economy. Available at https://www.pwc.com/gx/en/sustainability/assets/taking-control-in-a-circular-economy.pdf. (Accessed *18 May, 2020*).

PricewaterhouseCoopers (2017b). Setting your direction in the circular economy. Available at https://www.pwc.com/gx/en/services/sustainability/circular-business.html. (Accessed *19 May, 2020*).

Reichel, A., De Schoenmakere, M., Gillabel, J., Martin, J., & Hoogeveen, Y. (2016). Circular economy in Europe: Developing the knowledge base. *European Environment Agency Report, 2.*

Renwick, D. W., Redman, T., & Maguire, S. (2013). Green human resource management: A review and research agenda. *International Journal of Management Reviews, 15*(1), 1–14.

Rosen, C. M. (2001). Environmental strategy and competitive advantage: An introduction. *California Management Review, 43*(3), 9–16.

Rothenberg, S. (2003). Knowledge content and worker participation in environmental management at NUMMI. *Journal of Management Studies, 40*(7), 1783–1802.

Schneider, A. (2015). Reflexivity in sustainability accounting and management: Transcending the economic focus of corporate sustainability. *Journal of Business Ethics, 127*(3), 525–536.

Schuler, R. S., & Jackson, S. E. (1987). Linking competitive strategies with human resource management practices. *Academy of Management Perspectives*, *1*(3), 207–219.

Sempels, C., & Hoffmann, J. (2013). *Sustainable innovation strategy: Creating value in a world of finite resources*. UK: Springer.

Sikdar, S. (2019). Circular economy: Is there anything new in this concept? *Clean Technologies and Environmental Policy*, *21*, 1173–1175

Singh, J., & Ordoñez, I. (2016). Resource recovery from post-consumer waste: Important lessons for the upcoming circular economy. *Journal of Cleaner Production*, *134*, 342–353.

Sodhi, M. S. (2015). Conceptualizing social responsibility in operations via stakeholder resource-based view. *Production and Operations Management*, *24*(9), 1375–1389.

Soo Wee, Y., & Quazi, H. A. (2005). Development and validation of critical factors of environmental management. *Industrial Management & Data Systems*, *105*(1), 96–114. https://doi.org/10.1108/02635570510575216

Stahel, W. (2010). *The performance economy*. UK: Palgrave Macmillan.

Stahel, W. R. (2013). The business angle of a circular economy–higher competitiveness, higher resource security and material efficiency. *A new dynamic: Effective business in a circular economy*. Geneva: EMF.

Stahel, W. R. (2016). The circular economy. *Nature*, *531*(7595), 435–438.

Stock, T., & Seliger, G. (2011). Opportunities of sustainable manufacturing in Industry 4.0. *Procedia CIRP*, *40*, 536–541 (2016). In *13th Global Conference on Sustainable Manufacturing-Decoupling Growth from Resource Use*.

Su, B., Heshmati, A., Geng, Y., & Yu, X. (2013). A review of the circular economy in China: moving from rhetoric to implementation. *Journal of Cleaner Production*, *42*, 215–227.

Svensson, N., & Funck, E. K. (2019). Management control in circular economy. Exploring and theorizing the adaptation of management control to circular business models. *Journal of Cleaner Production*, *233*, 390–398. https://doi.org/10.1016/j.jclepro.2019.06.089

The Ellen MacArthur Foundation (EMF) (2013). Towards the Circular Economy. Available at https://www.ellenmacarthurfoundation.org/assets/downloads/publications/Ellen-MacArthur-Foundation-Towards-the-Circular-Economy-vol.1.pdf. (Accessed *18 May, 2020*).

The Ellen MacArthur Foundation (EMF) (2015). Towards a Circular Economy: Business rationale for an accelerated transition. Available at https://www.ellenmacarthurfoundation.org/publications/towards-a-circular-economy-business-rationale-for-an-accelerated-transition. (Accessed *18 May, 2020*).

Tse, T., Esposito, M., & Soufani, K. (2016). How businesses can support a circular economy. *Harvard Business Review*. Available at https://hbr.org/2016/02/how-businesses-can-support-a-circular-economy. (Accessed *18 May, 2020*).

Tukker, A. (2015). Product services for a resource-efficient and circular economy–a review. *Journal of Cleaner Production*, *97*, 76–91.

Ünal, E., Urbinati, A., & Chiaroni, D. (2018). Managerial practices for designing circular economy business models: The case of an Italian SME in the office supply industry. *Journal of Manufacturing Technology Management*, *30*(3), 561–589. https://doi.org/10.1108/JMTM-02-2018-0061

UNEP (2006). Circular Economy: An alternative for Economic Development. Paris, France: UNEP DTIE.

Urbinati, A., Chiaroni, D., & Chiesa, V. (2017). Towards a new taxonomy of circular economy business models. *Journal of Cleaner Production*, *168*, 487–498.

Uyarra, E., Edler, J., Garcia-Estevez, J., Georghiou, L., & Yeow, J. (2014). Barriers to innovation through public procurement: A supplier perspective. *Technovation*, *34*(10), 631–645.

Van Heerde, H. J., Gijsenberg, M. J., Dekimpe, M. G., & Steenkamp, J. B. E. (2013). Price and advertising effectiveness over the business cycle. *Journal of Marketing Research, 50*(2), 177–193.

von Bertalanffy, L. (1950). An outline of general system theory. *The British Journal for Philosophy of Science, 1*(2), 134–165.

Wehrmeyer, W. (1996). *Greening people: Human resource and environmental management.* New York: Greenleaf.

Wilkinson, A. (1992). The other side of quality: 'Soft' issues and the human resource dimension. *Total Quality Management, 3*(3), 323–330.

Winans, K., Kendall, A., & Deng, H. (2017). The history and current applications of the circular economy concept. *Renewable and Sustainable Energy Reviews, 68*, 825–833. https://doi.org/10.1016/j.rser.2016.09.123

World Business Council for Sustainable Development (2017). Policy makers, business leaders and the circular economy; driving solutions for sustainability. Available at https://www.wbcsd.org/Overview/News-Insights/Insights-from-the-President/Policymakers-business-leaders-and-the-circular-economy-driving-solutions-for-sustainability. (Accessed *19 May, 2020*).

World Commission on Environment & Development (1987). *Our common future.* New York: Oxford University Press.

World Economic Forum (2014). Towards the circular economy: Accelerating the scale-up across global supply-chain. Available at http://www3.weforum.org/docs/WEF_ENV_TowardsCircularEconomy_Report_2014.pdf (Accessed *17 May, 2020*)

Wright, P. M., McMahan, G. C., & McWilliams, A. (1994). Human resources and sustained competitive advantage: A resource-based perspective. *International Journal of Human Resource Management, 5*(2), 301–326.

Wu, H. Q., Shi, Y., Xia, Q., & Zhu, W. D. (2014). Effectiveness of the policy of circular economy in China: A DEA-based analysis for the period of 11th five-year-plan. *Resources, Conservation and Recycling, 83*, 163–175.

Ying, J., & Li-jun, Z. (2012). Study on green supply chain management based on circular economy. *Physics Procedia, 25*, 1682–1688.

Yuan, Z., Bi, J., & Moriguichi, Y. (2008). The circular economy: A new development strategy in China. *Journal of Industrial Ecology, 10*(1–2), 4–8.

Zhu, Q., & Geng, Y. (2013). Drivers and barriers of extended supply chain practices for energy saving and emission reduction among Chinese manufacturers. *Journal of Cleaner Production, 40*, 6–12.

2 Closing Loops, Easing Strains

Industry 4.0's Potential for Overcoming Challenges of Circularity in Manufacturing Environments

Lucas López-Manuel, Fernando León-Mateos, and Antonio Sartal
School of Economics and Business,
University of Vigo, Vigo, Spain

CONTENTS

2.1   INTRODUCTION

In recent decades, world markets have come under increasing competitive pressure that has created a huge concern in manufacturing industries to produce with lower costs, higher quality, and shorter delivery times (Sartal *et al.*, 2017). At the same time, the well-being levels in developed countries have favored a greater concern for social and environmental issues that require companies to leave behind management philosophies built exclusively on efficiency searches (Sartal *et al.*, 2020a). Business organizations need to internalize concern for the environment: from the design of their products to their industrialization and distribution, companies modify their behavior to comply with legal frameworks and respond to new 'green' demands from consumers (Kaswan and Rathi, 2020). Therefore, environmental sustainability is now a strategic imperative that must be added to firms' traditional objectives for profitability and efficiency (Quintás *et al.*, 2018; Arora et al., 2020).

As a result of the 19th century's unparalleled growth in economic output and population, the stress that humanity has long imposed on natural equilibria has reached alarming levels; regarding greenhouse emissions, for instance, the atmosphere's limited capacity to absorb the emissions released by today's carbon-based economies creates a hazard not only for the environment, but also for people's quality of life (Edenhofer *et al.*, 2015). At the same time, this growth model has reinforced the rising inequality between early industrialized and developing countries, expanding the gap between their populations' well-being.

Until the final decades of the 20th century, societies believed natural resources were an endless flow of raw materials and that the environment's regenerative capacity was able to compensate for all of the consequences of human action. Today, however, the facts have shown that the current paradigm of unlimited resource consumption is no longer acceptable. As people realize the importance environmental issues have in all organizational aspects of human life, it becomes necessary to include sustainability perspectives within economic, political and social stances. The publication of the Brundtland report (Brundtland Commission, 1987) pointed to the necessity of pursuing economic growth in a more sustainable way 'that meets the needs of the present without compromising future needs'. This report and subsequent treaties devised to battle the effects of climate change served to stimulate nations' environmental awareness; customers began to demand 'environmentally friendly' services and products, and companies began to notice new commercial opportunities (Defee *et al.*, 2009).

In this regard, achieving sustainable manufacturing practices in manufacturing industries is pivotal (Tan *et al.*, 2011; Shankar *et al.*, 2017; Alayón *et al.*, 2017). Sustainable manufacturing is a way of creating products through processes that ensure the conservation of energy and resources to ultimately minimize the aggregate negative impact on the environment (Moktadir *et al.*, 2018).

Since the First Industrial Revolution, production has been based on a linear economic system in which raw materials and resources were considered unlimited, following an 'obtain, use, and discard' approach. However, this model triggers non-renewable resource depletion, which results in severe economic and social impacts with wide ecological footprints in natural ecosystems (Momete, 2020).

These challenges along with the underlying limitations of a linear economy led to the appearance of the concept of a circular economy (CE) within the framework of sustainable manufacturing as a way to harmonize the ambitions of economic growth and the protection of the environment (Lieder and Rashid, 2016). The CE model, main objectives of which are to diminish the production of waste and pollution and the use of raw resources (e.g. fresh water), is having a great economic and ecological impact as it closes the 'loops' of flows in resource utilization (Ormazabal *et al.*, 2018; Bressanelli *et al.*, 2019; Sartal *et al.*, 2020c).

The CE places attention to the entire material life cycle, from the initial moment when things are designed, gathered from the soil, or sourced by upstream suppliers to their transformation by manufacturers. CE encompasses everything from distribution and sales to customer use and reuse. Indeed, the key aspect of circularity is building a strong reverse supply chain, through which worn products can be returned to manufacturing stages to be used as new inputs. Thus, materials can be further exploited while they circulate through a closed-loop supply chain system (Lieder and Rashid, 2016). As a CE considers societal, technological and environmental issues in terms of individual industrial processes, resource efficiency – with a particular focus on industrial waste generation – is sought to achieve better harmony and balance among society, the environment and the economy (Lieder and Rashid, 2016; Momete, 2020; Sartal *et al.*, 2020c).

The CE notion is gaining momentum as an ostensibly novel path toward sustainable development (Lieder and Rashid, 2016; Govindan and Hasanagic, 2018). Despite mounting attention and support, CEs have experienced little implementation thus far (Kirchherr *et al.*, 2018). Lieder and Rashid (2016) found that CE research has focused on waste generation, environmental impact and resource use while neglecting business and management viewpoints. This neglect, however, bears the risk of hindering CE implementation, as its advantages for industries are inexplicit. In their review of the literature on this subject, Vanner *et al.* (2014) identified several limiting factors for CE development: economic signs that do not foster innovation, the efficient use of resources, or pollution mitigation; lack of awareness and information; insufficient investment in technology; limited sustainable public incentives; and lower consumer and business acceptance.

In sum, the barriers that hamper the CE's development can be grouped into the following categories: financial, structural, operational, cultural, technological and governance (de Jesus and Mendonça, 2018; Govindan and Hasanagic, 2018; Araujo Galvão *et al.*, 2018). Among them, the fundamental barrier seems to be technological: the deployment of relevant technology is a precondition for the transition to a CE, as stated by much of the literature (e.g. Shahbazi *et al.*, 2016; Pheifer, 2017). When analyzing the barriers to implementation of a CE, de Jesus and Mendonça (2018) find that 'technical bottlenecks stand out as the perceived source of the greatest challenges' (p. 81). For their part, Kirchherr *et al.* (2018) affirm that existing literature calls technology the main barrier to a CE; 35% of studies raise this point, far more than for any other type of barriers.

Following this view, many practitioners and researchers have also raised the idea that 'Industry 4.0' (I4.0) offers tools and solutions to help solve this eco-efficiency challenge (Stock and Seliger, 2016; Sartal *et al.*, 2019; Horváth and Szabó, 2019). Sartal *et al.* (2020b) point out that the arrival of I4.0 can be a decisive facilitator for improving industrial process efficiency and optimizing CE models (Figure 2.1).

I4.0 is a multi-field notion that embodies the beginning of the Fourth Industrial Revolution, which was first introduced during the Hannover Fair event in 2011 (Glistau and Coello, 2018; Sartal *et al.*, 2019). The technologies this concept entails emphasize consistent digitization and information sharing by linking production units to achieve an interoperable environment. In doing so, new processes and

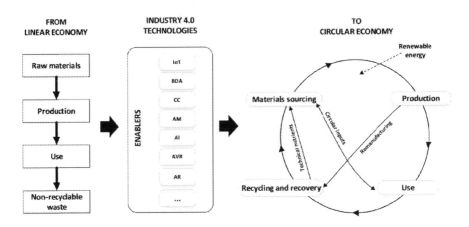

FIGURE 2.1   From linear economy to circular economy: the role of industry 4.0 technologies.

manufacturing technologies can be deployed, increasing sustainable and operational performance.

In accordance with these trends and considering the different I4.0 technologies as facilitators of CEs, our work explores how I4.0 can contribute to overcome the technological barriers to successfully establish CEs. To do this, we establish a theoretical framework based on the grouping of I4.0 technologies into three bundles: (i) connectivity and data analytics (CDA), (ii) autonomously supervised plant systems (ASPS) and (iii) virtual reality and optimization (VRO). This grouping allows us to thoroughly study the effects of I4.0 on a CE from a triple perspective: (i) the connectivity between agents and processes, (ii) the management revolution generated by the employment of new technologies in production processes and (iii) the further process and product optimization that virtual environments and forecasting techniques allow.

2.2   INDUSTRY 4.0 AND ITS UNDERPINNING TECHNOLOGIES

Throughout human history, the appearance of innovative technologies has resulted in astonishing changes in firm processes, structures and behaviors. The First Industrial Revolution saw the implementation of mechanized processes and steam engines in production facilities. From the mid-19th century onward, advances in electricity, communications and heavy industries, as well as in labor division, triggered the Second Industrial Revolution. Subsequently, the third industrial wave was set off in the second half of the 20th century, when information technologies (IT) and advanced electronics further improved communication systems (e.g. computers, the Internet) and process automation (Figure 2.2).Along with these technological evolutions, new organizational structures and strategies appeared. From the Fordist perspective of mass-production lines in the Second Industrial Revolution to Lean's waste reduction and process standardization in the third, the objectives of these strategies have always been to maximize efficiency while fulfilling customers' preferences (Sartal and Vázquez, 2017). In this regard, society is now experiencing a

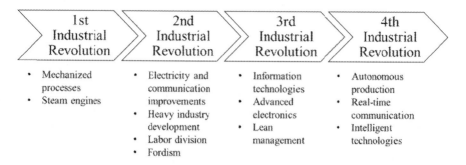

FIGURE 2.2 The four industrial revolutions and their changes.

transition between two stages which are driven by two different forces: customers' behavioral changes and technological development.

Consumers' preferences are moving toward products and services with greater differentiation as consumers are also increasingly concerned about environmental and social problems (Kaswan and Rathi, 2020). Therefore, firms need further adaptability. Production processes need to be quickly adapted to fulfill ever-changing customer demands, which are currently for ultra-personalized products and services (UPPS), to create value (Müller *et al.*, 2018; Sartal *et al.*, 2018; Torn and Vaneker, 2019). On the other hand, technology has reached a new stage; it is arguably considered to have broken from past trends and started a new one by itself: the Fourth Industrial Revolution (i.e. I4.0) was born from further IT development and its subsequent integration with computer-integrated manufacturing systems, usually referred to as cyber-physical systems (CPSs; Oztemel and Gursev, 2020; Mattos Nacimento *et al.*, 2019; Yao *et al.*, 2019). CPSs are born from the connection of physical items within virtual environments by using big data (BD), artificial intelligence (AI), cloud computing (CC) and the Internet of Things (IoT), which enable new business strategies (e.g. mass personalization, production decentralization), structures (e.g. modularity, dark factories) and processes (e.g. real-time monitoring, continuous optimization). With improved end-to-end communication (E2E), (i) information can be accessed and analyzed in real time; (ii) production capabilities, flexibility and efficiency will increase; and (iii) decision-making processes will be decentralized. Along with CPSs, innovative technologies such as additive manufacturing (AM), augmented virtual reality (AVR) and virtual simulation can be developed within these interconnected environments, with the potential to further improve operational performance (Barreto *et al.*, 2017; Wagner *et al.*, 2017; Leng *et al.*, 2019).

The interconnection of these systems across different processes and agents in manufacturing environments will enable the appearance of connected smart factories (CSFs) and change the paradigm of operational excellence practices (Liu and Xu, 2017; Oztemel and Gursev, 2020; Mattos Nacimento *et al.*, 2019). A CSF embodies a transversally connected network-based manufacturing system that fosters (i) improvements in process efficiency (higher energy efficiency, lower resource consumption and new manufacturing processes), (ii) the reduction of waste and replacement of raw materials, (iii) higher product-added value for customers (through increased product customization) and (iv) improved coordination while maintaining high-variety, low-cost and flexible production (Park, 2016; Torn and Vaneker, 2019; Ghobakhloo, 2020).

However, due to their characteristic interconnectivity, CSFs require principles from the deployment of communication and IT, highlighting the use of interoperability approaches (i.e. IoT, Internet of Services [IoS], and Internet of People [IoP]) and BDA to make the most of their benefits (Niesen *et al.*, 2016; Gilchrist, 2016; Hortelano *et al.*, 2017). Interoperability is the cornerstone for the components of a network – such as control systems, decision systems, human resources and intelligent equipment – to be connected and coordinated, sharing the data needed for optimal functioning (Tortorella and Fettermann, 2018; Zheng *et al.*, 2018). In these environments, the IoT, IoP and IoS become intertwined, triggering the appearance of multilayered communication flows called the Big Internet (TBI).

Nevertheless, this is not a constraint, as interoperability is already considered a necessary condition for I4.0 and CPS/CSF deployment (Li, 2018; Usuga Cadavid et al., 2020; Fragapane *et al.*, 2020). As a consequence, all of the technologies of CSFs must be equally developed to create smart manufacturing environments that are able to stimulate greater production, optimization and simulation capabilities (Liu and Xu, 2017; Yao *et al.*, 2019).

Since I4.0 is a paradigm involving an extensive range of different concepts and its technologies have most often been individually studied (Aceto *et al.*, 2019; Oztemel and Gursev, 2020), a division into three main bundles can be made to include the possibility for different outcomes, necessities and challenges within each set of CSF practices. An analysis of this kind would enable us to thoroughly study I4.0 as a trinity of fields with different natures, yet also as individual, and different technologies that are as follows: (i) swift connectivity between agents and processes of a network, (ii) new technologies in manufacturing processes that revolutionize production management practices and (iii) virtual environments and forecasting methods that allow further process and product optimization (Figure 2.3).

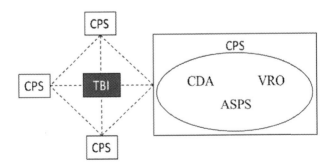

CPS: *cyber-physical systems.* CDA: *connectivity and data analytics.* VRO: *virtual reality and optimization.* ASPS: *autonomously supervised plant systems.* TBI: *the Big Internet*

FIGURE 2.3   The connected smart factory structure.

2.2.1 CONNECTIVITY AND DATA ANALYTICS (CDA)

Technologies such as TBI, BD, AI and Computer Systems (CS) enable interconnection between machines, human resources, materials and process controllers (Longo *et al.*, 2017; Frank *et al.*, 2019; Li *et al.*, 2020). While TBI (IoT, IoP, IoS) is the net that creates the cyber-environment, whereby information is conveyed. CSs are expected to store the BD gathered from manufacturing processes in remote systems. As CSs create platforms of computer networks with central hardware and distributed software, information can be accessed anywhere, providing services to different users (Rehman *et al.*, 2019; Majstorovic and Stojadinovic, 2020). Hence, cost and energy consumption are reduced, resulting in a better allocation of resources. This pillar is basic for the implementation of the other bundles since it will allow enhanced information flows, real-time analysis and process optimization, which are the foundations for a transverse system integration (Liu and Xu, 2017; Fragapane *et al.*, 2020; Usuga Cadavid et al., 2020).

Clearer information about inventories, machine conditions and transportation routes will bring transparency, surveillance and control to smart factories, resulting in efficiency improvements, reductions in time and reductions in waste and flaws in manufacturing processes (Buer *et al.*, 2018; Schroeder *et al.*, 2019; Ghobakhloo and Ng, 2019). In addition, these enhanced capabilities will provide enhanced troubleshooting and quicker problem-solving decisions (Li *et al.*, 2017; Kamble *et al.*, 2020). In this sense, human connectivity is also improved. Social intranets will be strengthened as the communication gap between the different layers of workers – employees, middle management and leadership – is reduced (de Zubielqui *et al.*, 2019).

As a consequence, the CDA pillar will unleash new competitive advantages, organizational capabilities, increased labor productivity and, ultimately, higher operational performance (Horvat *et al.*, 2019; Holmström *et al.*, 2019; Li *et al.*, 2020).

2.2.2 AUTONOMOUSLY SUPERVISED PLANT SYSTEMS (ASPS)

AM, automatic guided vehicles (AGVs) and autonomous robots and cobots (ARCs) are changing the way production processes are conceived. AM, also known as 3D printing, is a manufacturing process that 'uses computer-aided design to build objects layer by layer, as opposed to traditional manufacturing, which cuts, drills and grinds away unwanted excess from a solid piece of material', according to the ASTM International Committee F42 on AM technologies. This technology is expected to bring further savings in energy consumption and waste reduction, as well as flexibility in manufacturing environments (Friesike *et al.*, 2019; Ghobadian *et al.*, 2020). AGVs are means of transportation that, by using AI, allow intelligent, autonomous navigation to deliver materials across production environments. Their use will bring more flexibility to production lines, improving the material flow and increasing market responsiveness (Dias *et al.*, 2018; Fragapane *et al.*, 2020). Finally, ARCs are mechanized devices that arise from the technological possibilities established by AI and interconnectivity systems (e.g. IoT, CS) to improve processes. While autonomous robots make decisions without any operator by learning from surrounding environments, cobots are designed to help workers with their tasks, creating a symbiotic relationship between humans and machines (Thoben *et al.*, 2017; Ben-Ari and Mondada, 2018).

Aiming to reduce strain and improve efficiency across the production environment, these technologies unleash gains in resource usage (through smarter material planning, new allocation systems, or intelligent materials (Fragapane *et al.*, 2020; Ghobakhloo, 2020), and operational performance by (i) material flow optimization, (ii) better time to market, (iii) process optimization, (iv) resource efficiency, (v) reduction of waste, (vi) enhanced product innovation and quality, (vii) improved production capacity and reliability and (viii) lower inventory costs (de Sousa Jabbour *et al.*, 2018; Ghobadian *et al.*, 2020; Moreno *et al.*, 2020).

I4.0 can also further improve strategic adaptability. It can enable modular manufacturing environments that increase the agility, flexibility and adaptability of facilities, which can successfully adapt to changing consumer requirements (Kumar, 2018; Niaki *et al.*, 2019; Torn and Vaneker, 2019). This will eventually result in reduced time to market, production complexity and cost (Piran *et al.*, 2017; Ülkü and Hsuan, 2017; Åkerman *et al.*, 2018).

These technologies pave the way for not only operational benefits, but also for employee security, as they reduce human intervention. Intelligent collaborative robots (cobots), 3D printers and AGVs show higher risk assessment and hazard identification capabilities; they understand the surrounding world better, make fewer errors and keep human colleagues safer (Schou *et al.*, 2018; Bragança *et al.*, 2019; Fragapane *et al.*, 2020). In addition, the resulting automation and process simplification, along with data-improved decision-making, could also boost human resource efficiency.

2.2.3   VIRTUAL REALITY AND OPTIMIZATION (VRO)

Virtual reality (VR) is a virtual representation of the world and is created by computers that compile information gathered from real environments (Guizzi *et al.*, 2020). As a result, a virtual environment – another alternative reality – appears, becoming the basic structure for simulation tools to be implemented. On one hand, the use of digital twins – a virtual replica of an item, process, or service – allows for simulation-based predictions and manufacturing optimizations, as well as control of product life cycle or process performance (Qi and Tao, 2018; Tao *et al.*, 2018; Leng *et al.*, 2019). On the other hand, AVR – a mix between real and virtual environments that enhances human perception and operability through the use of digital information treatment systems such as Head Up Displays (HUD) or interactive overlays – can be used to effectively train workers. AVR is especially valuable in hazardous, sensitive tasks (Sun *et al.*, 2019; Guizzi *et al.*, 2020). It also can be used for matching vacancies with applicants, optimizing HR recruiting processes (Stone *et al.*, 2018).

As a result, this set of practices will enable better forecasting abilities for businesses to (i) deal efficiently with environmental uncertainties, (ii) control processes of change further and (iii) adapt business models to changing conditions (Yin *et al.*, 2018; Kusiak, 2018). Therefore, firms can try new configurations while avoiding real trial-and-error approaches that could be harmful or risky.

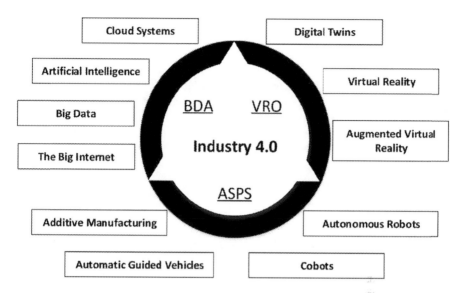

FIGURE 2.4   I4.0 basic technologies under the Industry 4.0 umbrella.

Finally, if the CSF concept is applied across the entire supply chain, the latter will be transformed into a digital supply network (DSN). A DSN enables a horizontally and vertically unified supply network where all agents are linked and integrated, able to interconnect with each other worldwide and in real time (Kache and Seuring, 2017; Ardito *et al.*, 2019). The 'smartness' of the manufacturing process will spill over into the rest of the supply chain, making it smarter as well (see Figure 2.4).

2.3   INDUSTRY 4.0 OPPORTUNITIES TO ESTABLISH CE ENVIRONMENTS

Since linearity in economy models cannot be sustainable because it creates large amounts of waste in addition to depleting the available natural resources (Genovese *et al.*, 2017), the CE is arising as a strategy for achieving sustainable development that considers production processes such as the supply chain's consumption activities (Sehnem *et al.*, 2019). The CE proposes a cradle-to-cradle (C2C) production and consumption approach, promoting a new industry concept in which everything can be reutilized, whether the product is reintroduced to the industry as a technical nutrient that can be recycled over and over again or returned to the earth as a nontoxic biological nutrient. The objective is that the value of resources (water, energy, etc.), materials and products remains in the economy for as long as possible, diminishing waste creation. In this way, the CE proposes to close the life cycle of energy, water, materials, products, services and waste. According to Stahel and Reday-Mulvey (1981), a closed-loop CE favors reusing, repairing and remanufacturing over the manufacturing of new goods and has a positive impact in terms of waste prevention, resource savings, job creation and economic competitiveness.

To do this, by-products and services are used according to the 6R principles: reuse, reduce, recycle, redesign, remanufacture and repair operations for products (Lüdeke-Freund *et al.*, 2019). The CE is an uninterrupted cycle of positive development that optimizes resource performance, preserves and enhances natural capital, and minimizes system risks by managing renewable flows and finite stocks (Ellen MacArthur Foundation & McKinsey Center for Business and Environment, 2015).

Next, I4.0 technologies will be related to the cycle they affect, either the biological one or the technical one. Then, following a transversal approach, the synergies between I4.0 and CE will be identified according to those arising from products, companies or supply chains (Figure 2.5).

2.3.1 OPPORTUNITIES FOR BIOLOGICAL AND TECHNICAL CYCLE IMPROVEMENTS

Circular production systems became popular, thanks to the C2C notion defined by McDonough and Braungart (2002) in their seminal work on this theory. Both C2C and CE focus on remodeling the production cycle through evolution of an industrial

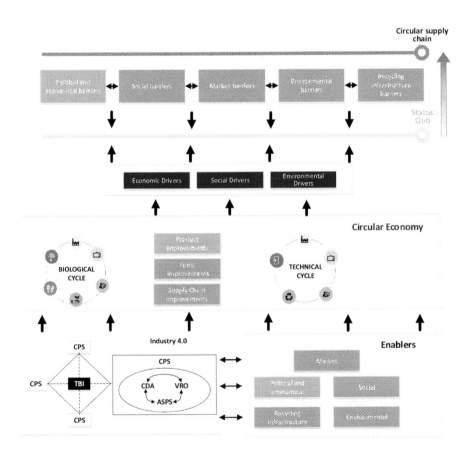

FIGURE 2.5   Opportunities for biological and technical cycle improvements.

system valuable for both ecological and human well-being (Lieder and Rashid, 2016). The C2C design regards all materials as nutrients that eradicate the notion of waste and employs the productive yet also safe processes of the nature's biological metabolism as a guide for developing a flow in the technical metabolism of industrial materials (Ellen MacArthur Foundation & McKinsey Center for Business and Environment, 2015). All industrial processes and products must be conceived in such a way that they allow the continuous stream of nutrients to flow through one of the following metabolisms (cycles): technical metabolism or biological metabolism (Braungart *et al.*, 2006; MacArthur *et al.*, 2015; Murray *et al.*, 2017).

In the biological cycle, consumer products (used until the end of their life cycle, for example, when a high degree of physical degradation or abrasion appears) must be organic and biodegradable materials that can 'decompose and become food for plants and animals as well as nutrients for soils' (McDonough and Braungart, 2002, p. 91). This cycle aims to regenerate ecosystems by promoting renewable materials and energy, lessening the consumption of natural resources and reintroducing waste in manufacturing as an input.

In the technical cycle, service products, or technical nutrients, are manmade materials that should flow within a closed-loop manufacturing, recovery and reuse system (technical metabolism), without contaminating the biosphere. These technical nutrients (composed of nontoxic durable composite materials) are expected to 'go back to upstream manufacturing cycles to supply high-quality raw materials for new products' (McDonough and Braungart, 2002, p. 91). Thus, these mineral or synthetic materials would remain within a closed-loop manufacturing system without experiencing any loss of performance or quality (Brennan *et al.*, 2015). The technical cycle underlines the enlargement of a product's life through reuse, repair, remanufacturing and recycling.

In this context, I4.0 technologies can close existing gaps in supply chains by improving efficiency, reuse of materials and, consequently, circular flows; CDA, ASPS and VRO technologies can address the lack of information and coordination between companies, enable cleaner technologies, or reduce uncertainty about operational performance, the main challenges of a CE (Su *et al.*, 2013; de Sousa Jabbour *et al.*, 2018b).

Regarding the biological cycle, CPSs can be useful at the time of design with the installation of sensors or chips that inform users of the materials and components a product is made of and how these leftovers can be recycled and disassembled at the end of their useful lives. This is what has been named the 'product passport' (European Commission, 2013). In this sense, many academics regard CPSs as a way to allow better product life cycle management or the development of new services (Caggiano, 2018).

ASPS, through AM, can allow the exploitation of biomaterials, which can be converted easily into biological nutrients at the end of the product life cycle (van Wijk and van Wijk, 2015; Voet *et al.*, 2018). AM could support the life cycle management of products and processes (Ribeiro *et al.*, 2020). BDA also contributes to collecting or managing product life cycle data (Ren *et al.*, 2019), which can be used to improve the planning of disassembly sequences (Marconi *et al.*, 2019). By reflecting on recycling issues from the product design stage (Lin, 2018), TBI allows data collection throughout the Life Cycle Performance (LCP) framework. Furthermore, the use of the IoT can extend the product life cycle while allowing for new strategies of waste management (Esmaeilian *et al.*, 2018). In this case, VRO can improve efficiency in the exploitation of natural resources – for example, by calculating eco-efficiency indices (Rönnlund *et al.*, 2016). Furthermore, simulation can be used to assess environmental impacts (Deschamps *et al.*, 2018).

Regarding the technical cycle, logistics and reverse logistics processes could be improved through the IoT because sensors, barcodes, or RFID tags can optimize logistic routes – through, for example, information sharing policies – or track post-consumer products and packaging, among other initiatives (e.g. Groba *et al.*, 2020). Knowing the location of these waste materials would facilitate their reentry into the process as new inputs for production. The IoT can also contribute both to the optimization of supply chain management performance (Ben-Daya *et al.*, 2019) and to the design of remanufacturing processes (French *et al.*, 2017). As in the biological cycle, CPS allows the product to be monitored throughout its life cycle, which allows for a better understanding of how it can be dismantled and recycled at the end of its life.

BDA, through CC, could help organizations develop circular business models that draw upon recycled materials (Mattos Nascimento *et al.*, 2018), allowing companies to find buyers for restored or reused components (MacArthur and Waughray, 2016). For its part, TBI serves to connect all members of the value chain, allowing people to find industrial symbiosis (Song *et al.*, 2017). It also facilitates the development of automated approaches that can assess the potential value of secondary materials (Davis *et al.*, 2017; Jose and Ramakrishna, 2018).

The characteristics of AM lead to the reduced use of material. In addition, this technology allows the recycling of minor amounts of waste thanks to the 3D printers' movability. AM also favors the updating of current recycling processes (Clemon and Zohdi, 2018; Sauerwein and Doubrovski, 2018), thanks to the ability offered by 3D printers to use granular compounds of plastic and granulated metallic materials.

VROs, for their part, allow the establishment of supply chain management performance optimization methods – for example, through digital twins (Ivanov *et al.*, 2019) or material flow modeling (Schäfers and Walther, 2017). The simulation also serves as a complementary tool for recycling since it can be useful for calculating recycling performance indices (van Schaik and Reuter, 2016).

2.3.2 OPPORTUNITIES FOR PRODUCT IMPROVEMENTS

In a linear economy, a product is the result of raw materials that have undergone several manufacturing processes before being delivered to the customer. Eventually, when consumers see it as having no value, they dispose of it, and the product ends up in dumps, is burned or, in the best of cases, is partially recycled. Conversely, the CE expects products to be valuable the longest, either as a final good for customers or as an input for remanufacturing (Geissdoerfer *et al.*, 2017; Genovese *et al.*, 2017). Thus, the overarching objective of circularity can be accomplished by decreasing the amount of fresh raw materials introduced into the system, which eases environmental strain on ecosystems. In this approach, I4.0 can be useful for product design, end-of-life (EOL) recovery and mass customization (Pardo *et al.*, 2020; Sartal *et al.*, 2020a).

Through TBI, data can be gathered throughout product life and customer use. The possibility of obtaining real-time product information allows manufacturers to assess environmental KPIs' performance and improve life cycle assessment (Gbededo *et al.*, 2018; Jose and Ramakrishna, 2018). Once this information is obtained, BDA technologies can be used to evaluate product performance, spot possible flaws, reduce design costs, and further enhance green behaviors, such as the use of easy-to-recycle materials or longer product lifetimes (Kiel *et al.*, 2017; Cao *et al.*, 2019; Li *et al.*, 2020). Furthermore, following the research of Tiwari and Khan (2020), CPS and data-gathering capabilities have proven useful for increasing the sustainability commitment. This is done through products' material disclosures; as environmental performance monitoring can be enhanced. On the other hand, VRO techniques can also enhance the potential offered by BDA, helping forecast product life cycle or product performance (Qi and Tao, 2018; Leng *et al.*, 2019).

Once product cycles are understood, their design can be made greener, improving products' EOL; when the need for EOL recovery is considered, product design can draw attention to the ease of reuse, remanufacture or recycling. In this sense, Joshi and Gupta (2019) created an 'Advanced-Remanufacturing-To-Order-Disassembly-To-Order system' to assess EOL designs. Following this kind of framework, information gathered throughout product life can help decision-making processes at this stage; if the materials are useful, they will be reintroduced in new manufacturing processes. Otherwise, they will go on to disposal facilities.

Also, ASPS can increase products' suitability for undergoing these processes. In this regard, AM can diminish the complexity of products while using new materials with increased capabilities for reutilization (Jose and Ramakrishna, 2018; Sartal *et al.*, 2019; Niaki *et al.*, 2019).

BDA, VRO and ASPS play significant roles in products' mass customization, as understanding and fulfilling customer preferences are the keys for developing new products to better match those preferences. The opportunity to analyze customer and product data provides possibilities for improving decision-making processes regarding customer and product decisions, increasing energetic and material resource efficiency (Fosso Wamba *et al.*, 2015; Tao *et al.*, 2018). As green customer concerns increase, companies can offer greater value, resulting in the possibility of gaining green competitive advantages. Regarding products, AM could be the optimal solution to fulfilling ever-changing customer desires, as product designs can be modified just before manufacturing (Dalenogare *et al.*, 2018; Sartal *et al.*, 2019; Ghobadian *et al.*, 2020).

2.3.3 OPPORTUNITIES FOR FIRM IMPROVEMENTS

Firms play a key role in establishing a CE. They not only have to enhance their process efficiency to employ fewer natural resources, but also are the agents that reintroduce disposal into manufacturing processes, closing the gaps and seizing the remaining value of waste. Therefore, two areas can be identified where firms need to thrive to effectively turn linear economies into circular ones. First, they need to achieve greater energy efficiency and lower resource consumption. Second, they have to manage waste retrieval, recycling and remanufacturing processes, which are critical points in closing manufacturing loops. All this must be done while upholding profitability, operational performance and customer satisfaction. In this regard, I4.0 technologies can unleash further improvements across firm processes, enabling a vertical integration that helps to successfully establish CEs (Sartal *et al.*, 2020b).

Greater efficiency and lower resource consumption can be achieved through the use of several I4.0 technologies. Gathering information across the different processes allows for greater control and further optimization of existing resources (Cao *et al.*, 2019; Li *et al.*, 2020). Better production planning and control, improved decision-making and enhanced learning processes are key factors of TBI and BDA technologies for instituting greener practices across manufacturing environments (Gbededo *et al.*, 2018; de Sousa Jabbour *et al.*, 2018b; Dubey *et al.*, 2019). On the other hand, the use of CEs results in improved resource allocation and data availability and reduces the need for investing in expensive, high-performance data systems, as data storage can be provided by third parties (Liu and Xu, 2017; Fragapane *et al.*, 2020). A firm is now able to analyze information flows arriving from both the supplier and the customer sides, which maximizes its involvement within the supply chain practices. Therefore, the combination of CDA will create new added value from data, enhancing flexibility, production management and, ultimately, mass customization (Kamble *et al.*, 2018; Tao *et al.*, 2018).

However, this is only possible if advanced ASPS with the ability to gather process information are also established. While AGVs smooth shop-floor logistics, ARCs provide in-time system information, remote diagnosis and control, self-organization and continuous optimization – fundamental processes for assessing resource use and waste creation (Ben-Ari and Mondada, 2018; Oztemel and Gursev, 2020; Fragapane *et al.*, 2020). Thus, the application of CDA technologies and ASPSs along with VRO can result in higher worker responsiveness and improvements in operational performance. Arguably, it is expected that lean manufacturing practices (e.g. Kaizen, Kanban, Just in Time – JIT) and I4.0 can create synergies to further enhance sustainability performance as efficiency, waste reduction and continuous improvement are common objectives (Gbededo *et al.*, 2018; Quenehen *et al.*, 2019; Bittencourt *et al.*, 2019) and the positive relationship between lean manufacturing and sustainability has been widely proved (e.g., Garza-Reyes, 2015; Sartal *et al.*, 2018; Sartal *et al.*, 2020a). Indeed, the performance of lean practices can even be heightened by I4.0, as its technologies provide greater production modularity and flexibility (which address mass customization issues) as well as eased, safer worker tasks (Koch *et al.*, 2017; Åkerman *et al.*, 2018; Torn and Vaneker, 2019). Following this view, the work of Sanders *et al.* (2016) presents 10 integration challenges lean manufacturing experiences that can be overcome with I4.0 implementation. Within ASPS, and regarding waste reduction in manufacturing, AM plays a special role, as the layer-by-layer manufacturing process eliminates waste instead of reducing it. In addition, AM can lower setup times and offer energy savings with subsequent lower indirect CO_2 emissions (Sartal *et al.*, 2018; Ghobadian *et al.*, 2020).

Thus, the deployment of CPSs and subsequent creation of CSFs is twofold, considering CE needs. First, they help achieve higher levels of resource efficiency, energy savings and waste reduction, and then increase operational performance through the use of I4.0 and lean philosophies (Quenehen *et al.*, 2019; Horvat *et al.*, 2019;

Fragapane *et al.*, 2020). Second, within CSMs, recycling and remanufacturing processes can be carried out more easily.

While CDA and CRO can help in introducing or improving reverse logistics, ASPS can provide a safer, more optimal way to develop product reconditioning processes (Alvarez de los Mozos and Renteria, 2017; Joshi and Gupta, 2019). The recycling of hazardous, delicate waste (e.g. lithium-ion batteries, e-waste) can be carried out with the help of automated systems that barely require human intervention while seizing the potential remaining value of the disposal.

These new manufacturing practices can bring about competitive advantages for firms, related not only to the improvement of operational excellence practices, but also to customers' perceptions that the company has integrated green values into its core values (Tao *et al.*, 2018; Cao *et al.*, 2019).

However, product and firm transformations toward more sustainable frameworks are not sufficient. For a CE to be fully deployed, horizontal integration across the entire supply chain must also occur.

2.3.4 OPPORTUNITIES FOR SUPPLY CHAIN IMPROVEMENTS

In recent years, increasing environmental concerns have driven the focus within supply chains to introducing sustainability pressures among their agents from product design to delivery, creating the concept of a green supply chain (Srivastava, 2007). However, greater coordination between firms is required to actually achieve further emission and waste reduction – fundamental conditions for establishing a CE – as well as dynamic reverse logistics for remanufacturing processes (Kamble *et al.*, 2018; Cao *et al.*, 2019). When improving connectivity between agents, I4.0 technologies can unleash several opportunities to overcome this challenge.

The use of CDA technologies, not only at the firm level but across supply chains, allows for multidirectional information exchanges, from the first tier of suppliers downward and vice versa. These can extend the benefits of information analysis. Transparency between firms and their response times are increased as connections become smoothed; resource efficiency, decision-making and logistics become optimized across supply chains as firms increase their sustainable performance (de Sousa Jabbour *et al.*, 2018; Dubey *et al.*, 2019). This improvement, highly regarded by customers with sustainability concerns, results in additional created value for firms. In addition, further supply chain integration will result in higher operational performance, improving the competitive advantages of the whole supply chain that can underpin the transition to greener supply chain practices (Lee and Lee, 2015; Bruque-Cámara *et al.*, 2016; Dalenogare *et al.*, 2018).

As Li *et al.* (2020) reveal that the success of digital technologies lies in the establishment of DSNs. Digital technologies must overcome firms' boundaries and integrate both downstream and upstream supply chain agents to maximize the value of I4.0 effects in highly competitive environments. DSNs thus become a cornerstone of improving economic and sustainable performance, as they are the mediating mechanism between digital technologies and environmental performance.

These changes will radically alter the way supply chains behave. With the appearance of CPSs and AGVs, flows and exchanges are enhanced, not only at the firm level but across the entire supply chain. Notwithstanding this, AM will be the revolutionary technology that sets out disruptive changes in logistics since it entails lighter and fewer deliveries, smaller batches and reductions in manufacturing steps (Sartal *et al.*, 2018; Niaki *et al.*, 2019). Indeed, as Ghobadian *et al.* (2020) highlight, AM alters the backbone of traditional supply chains; jobbing or batch processes are now replaced by 3D printers that do not cut, bash or weld materials. As a consequence, JIT practices will also be adjusted as firms can manufacture products on demand by loading only the specified design in the printer's software. Product stock will not be required anymore while inventories move toward storing raw material, shortening supply chains, reducing time to market and reducing the need for transportation (Table 2.1).

TABLE 2.1
Corollary: I4.0 Technologies in CE Environments: Potential Improvements and Key Factors

Bundle	Technology	CE Improvements	Barriers/Enablers
Connection and Data Analytics (CDA) Liu and Xu, 2017 Tao *et al.*, 2018 Kamble *et al.*, 2018 Frank *et al.*, 2019 Cao *et al.*, 2019 Li *et al.*, 2020 De Sousa Jabbour *et al.*, 2018	**The Big Internet (TBI)** **Big Data and Analytics (BDA)** **Cloud Systems (CS)**	• Production flexibility • Material use efficiency • Information processing efficiency • Decision-making • Emission and waste reduction • Identification of environmental and operational tradeoffs • Eco-practices: life cycle assessment, eco-design, data monitoring and evaluation of KPIs, products' end-of-life recovery	• Collaboration across supply chains • Organizational change • Privacy and security • Workforce's digital capabilities • Technological infrastructure and firm strategies • Need for investments
Autonomously Supervised Plant Systems (ASPS) Sartal *et al.*, 2018 Ghobadian *et al.*, 2020 Oztemel and Gursev, 2020 Dalenogare *et al.*, 2018 Quenehen *et al.*, 2019 Fragapane *et al.*, 2020	**Automatic Guided Vehicles (AGV)** **Autonomous Robots and Cobots (ARC)** **Additive Manufacturing (AM)**	• Material use efficiency • Energy consumption efficiency • Production flexibility and modularity • Production efficiency • Mass customization • Emission and waste reduction • Continuous improvement and self-organization • Flaw spotting and monitoring • Recycling, disassembly, remanufacturing and reuse • Emission and waste reduction • Product performance and durability • Integration between supply chain agents	• Lean principles • Organizational changes • Risk of failure due to their still-formative stage • Social, operational and economic disruptions • Employee reluctance • Need for CDA technologies

(Continued)

TABLE 2.1 *(Continued)*
Corollary: I4.0 Technologies in CE Environments: Potential Improvements and Key Factors

Bundle	Technology	CE Improvements	Barriers/Enablers
Virtual Reality and Optimization (VRO) Qi and Tao, 2018 Kusiak, 2018 Gbededo *et al.*, 2018 Oztemel and Gursev, 2020 Leng *et al.*, 2019	Virtual Reality (VR) Augmented Virtual Reality (AVR) Digital Twins (DT)	• Green design of products and production systems • Production efficiency • Decision-making and forecasting capabilities • Product life cycle assessment: quality and performance • Eco-design	• Need for investments • Workforce's digital capabilities • Technological advancement and infrastructure • Risk of failure due to their still-nascent development stage • Need for CDA and ASPS technologies

REFERENCES

Aceto, G., Persico, V., & Pescape, A. (2019). A survey on information and communication technologies for Industry 4.0: State-of-the-art, taxonomies, perspectives, and challenges. *IEEE Communications Surveys & Tutorials*, *21*(4), 3467–3501.

Agyemang, M., Kusi-Sarpong, S., Khan, S., Mani, V., Rehman, S., & Kusi-Sarpong, H. (2019). Drivers and barriers to circular economy implementation. *Management Decision*, *57*(4), 971–994.

Åkerman, M. F.-B., Halvordsson, E., & Stahre, J. (2018). Modularized as sembly system: A digital innovation hub for the Swedish smart industry. *Manufacturing Letters*, *15*(1), 143–146.

Alayón, C., Säfsten, K., & Johansson, G. (2017). Conceptual sustainable production principles in practice: Do they reflect what companies do? *Journal of Cleaner Production*, *141*, 693–701. doi:https://doi.org/10.1016/j.jclepro.2016.09.079.

Alvarez de los Mozos, E., & Renteria, A. (2017). Collaborative robots in e-waste management. *Procedia Manufacturing*, *11*, 55–62.

Araujo Galvão, G. D., de Nadae, J., Clemente, D. H., Chinen, G., & de Carvalho, M. M. (2018). Circular economy: Overview of barriers. *Procedia CIRP*, *73*, 79–85.

Ardito, L., Petruzzelli, A., Panniello, U., & Garavelli, A. (2019). Towards Industry 4.0: Mapping digital technologies for supply chain management-marketing integration. *Business Process Management Journal*, *25*(2), 323–346.

Arora, P., Hora, M., Singhal, V., & Subramanian, R. (2020). When do appointments of corporate sustainability executives affect shareholder value? *Journal of Operations Management*, *66*(4), 464–487.

ASTM International. (2017). *Helping our world work better*. ASTM INTERNATIONAL.

Barreto, L., Amaral, A., & Pereira, T. (2017). Industry 4.0 implications in logistics: An overview. *Procedia Manufacturing*, *13*, 1245–1252.

Ben-Ari, M., & Mondada, F. (2018). Robots and Their Applications. In M. Ben-Ari, & F. Mondada (Eds.), *Elements of robotics* (pp. 1–20). Cham, Switzerland: Springer International Publishing.

Ben-Daya, M., Hassini, E., & Bahroun, Z. (2019). Internet of things and supply chain management: A literature review. *International Journal of Production Research*, *57*(15–16), 4719–4742.

Bittencourt, V., Saldanha, F., Alves A. C., & Leão, C. (2019). Contributions of Lean Thinking Principles to Foster Industry 4.0 and Sustainable Development Goals. In A. Alves, F. J. Kahlen, S. Flumerfelt, & A. Siriban-Manalang (Eds.), *Lean engineering for global development* (pp. 129–159). Cham: Springer.

Bragança, S., Costa, E., Castellucci, I., & Arezes, P. (2019). A brief overview of the use of collaborative robots in Industry 4.0: Human role and safety. In P. Arezes, J. Baptista, M. Barroso, P. Carneiro, P. Cordeiro, N. Costa, … G. Perestrelo (Eds.), *Occupational and environmental safety and health* (pp. 641–650). Cham, Switzerland: Springer.

Braungart, M., McDonough, W., & Bollinger, A. (2006). Cradle-to-cradle design: Creating healthy emissions – a strategy for eco-effective product and system design. *Journal of Cleaner Production, 15*(13–14), 1337–1348.

Brennan, G., Tennant, M., & Blomsma, F. (2015). Business and production solutions: Closing the loop. I. In K. H., & E. Shoreman-Ouimet (Eds.), *Sustainability: Key issues* (pp. 219–239). New York (USA): Routledge.

Bressanelli, G., Perona, M., & Saccani, N. (2019). Challenges in supply chain redesign for the circular economy: A literature review and a multiple case study. *International Journal of Production Research, 57*(23), 7395–7422.

Brundtland Commission (1987). *Our common future.* Oxford (UK): Oxford University Press.

Bruque-Cámara, S., Moyano-Fuentes, J., & Maqueira-Marín, J. (2016). Supply chain integration through community cloud: Effects on operational performance. *Journal of Purchasing and Supply Management, 22*(2), 141–153.

Buer, S.-V., Strandhagen, J., & Chan, F. (2018). The link between Industry 4.0 and lean manufacturing: Mapping current research and establishing a research agenda. *International Journal of Production Research, 56*(8), 2924–2940.

Caggiano, A. (2018). Cloud-based manufacturing process monitoring for smart diagnosis services. *International Journal of Computer Integrated Manufacturing, 31*(7) 1–12.

Cao, G., Duan, Y., & Cadden, T. (2019). The link between information processing capability and competitive advantage mediated through decision-making effectiveness. *International Journal of Information Management, 44*, 121–131.

Clemon, L. M., & Zohdi, T. (2018). On the tolerable limits of granulated recycled material additives to maintain structural integrity. *Construction and Building Materials, 167*, 846–852.

Dalenogare, L., Benitez, G., Ayala, N., & Frank, A. (2018). The expected contribution of Industry 4.0 technologies for industrial performance. *International Journal of Production Economics, 204*, 383–394.

Davis, C. B., Aid, G., & Zhu, B. (2017). Secondary resources in the bio-based economy: A computer assisted survey of value pathways in academic literature. *Waste and Biomass Valorization, 8*(7), 2229–2246.

de Jesus, A., & Mendonça, S. (2018). Lost in transition? Drivers and barriers in the eco-innovation road to the circular economy. *Ecological economics, 145*, 75–89.

de Sousa Jabbour, A., Jabbour, C., Foropon, C., & Filho, M. (2018). When titans meet – Can Industry 4.0 revolutionise the environmentally-sustainable manufacturing wave? The role of critical success factors. *Technological Forecasting and Social Change, 132*, 18–25.

de Sousa Jabbour, A., Jabbour, C., Godinho Filho, M., & Roubaud, D. (2018b). Industry 4.0 and the circular economy: A proposed research agenda and original roadmap for sustainable operations. *Annals of Operations Research, 270*(1–2), 273–286.

de Zubielqui, G., Fryges, H., & Jones, J. (2019). Social media, open innovation & HRM implications for performance. *Technological ForecastIng and Social Change, 144*(1), 334–347.

Defee, C., Esper, T., & Mollenkopf, D. (2009). Leveraging closed-loop orientation and leadership for environmental sustainability. *Supply Chain Management, 14*(2), 87–98.

Deschamps, J., Simon, B., Tagnit-Hamou, A., & Amor, B. (2018). Is open-loop recycling the lowest preference in a circular economy? Answering through LCA of glass powder in concrete. *Journal of Cleaner Production, 185,* 14–22.

Dias, L., de Oliveira Silva, R., da Silva Emanuel, P., Filho, A., & Bento, R. (2018). Application of the fuzzy logic for the development of autommnomous robot with obstacles deviation. *International Journal of Control, Automation and Systems, 16*(2), 823–833.

Dubey, R., Gunasekaran, A., Childe, S., Papadopoulos, T., Luo, Z., Wamba, S., & Roubaud, D. (2019). Can big data and predictive analytics improve social and environmental sustainability? *Technological Forecasting and Social Change, 144,* 534–545.

Edenhofer, O., Pichs-Madruga, R., Sokona, Y., Farahani, E., Kadner, S., Seyboth, K., ... et al. (2015). IPCC, 2014: Summary for Policymakers. *In Climate Change 2014: Synthesis Report. Contribution of Working Groups I, II and III to the Fifth Assessment Report of the Intergovernmental Panel on Climate Change.* Cambridge, United Kingdom and New York, NY, USA: Cambridge University Press.

Ellen MacArthur Foundation, & McKinsey Center for Business and Environment (2015). *Growth within: A circular economy vision for a competitive Europe.* Cowes, UK: *Ellen MacArthur Foundation.*

Esmaeilian, B., Wang, B., Lewis, K., Duarte, F., Ratti, C., & Behdad, S. (2018). The future of waste management in smart and sustainable cities: A review and concept paper. *Waste Management, 81,* 177–195.

European Commission (2013). *European resource efficiency platform pushes for product passports.* Retrieved May 13, 2020, from http://ec.europa.eu/environment/ecoap/about-eco-innovation/policies-matters/eu/20130708_european-resource-efficiency-platform-pushes-for-product-passports_e

Fosso Wamba, S., Akter, S., Edwards, A., Chopin, G. & Gnanzou, D (2015). How 'big data' can make big impact: Findings from a systematic review and a longitudinal case study. *International Journal of Production Economics, 165,* 234–246.

Fragapane, G., Ivanov, D., Peron, M., Sgarbossa, F., & Strandhagen, J. (2020). Increasing flexibility and productivity in Industry 4.0 production networks with autonomous mobile robots and smart intralogistics. Annals of Operations Research. https://doi.org/10.1007/s10479-020-03526-7

Frank, A., Dalenogare, L., & Ayala, N. (2019). Industry 4.0 technologies: Implementation patterns in manufacturing companies. *International Journal of Production Economics, 210,* 15–26.

French, R., Michalis, B., & Marin-Reyes, H. (2017). Intelligent Sensing for Robotic Re-Manufacturing in Aerospace – An Industry 4.0 Design Based Prototype. *IRIS 2017 – 5th IEEE International Symposium on Robotics and Intelligent Sensors,* 272–277.

Friesike, S., Flath, C. M., Wirth, M., & Thiesse, F. (2019). Creativity and productivity in product design for additive manufacturing: Mechanisms and platform outcomes of remixing. *Journal of Operations Management, 65*(8), 735–752

Garza-Reyes, J. (2015). Lean and green – a systematic review of the state of the art literature. *Journal of Cleaner Production, 102,* 18–29.

Gbededo, M., Liyanage, K., & Garza-Reyes, J. (2018). Towards a life cycle sustainability analysis: A systematic review of approaches to sustainable manufacturing. *Journal of Cleaner Production, 184,* 1002–1015.

Geissdoerfer, M., Savaget, P., Bocken, N., & Hultink, E. (2017). The circular economy – a new sustainability paradigm? *Journal of Cleaner Production, 143*(Suplemment C), 757–768.

Genovese, A., Acquaye, A., Figueroa, A., & Koh, S. (2017). Sustainable supply chain management and the transition towards a circular economy: Evidence and some applications. *Omega, 66*(Part B), 344–357.

Ghobadian, A., Talavera, I., Bhattacharya, A., Kumar, V., Garza-Reyes, J., & O'Regan, N. (2020). Examining legitimatisation of additive manufacturing in the interplay between innovation, lean manufacturing and sustainability. *International Journal of Production Economics, 219*, 457–468.

Ghobakhloo, M. (2020). Industry 4.0, digitization, and opportunities for sustainability. *Journal of Cleaner Production, 252*, 119869.

Ghobakhloo, M., & Ng, T. (2019). Adoption of digital technologies of smart manufacturing in SMEs. *Journal of Industrial Information Integration, 16*, 100107.

Gilchrist, A. (2016). *Industry 4.0: The industrial internet of things.* Heidelberg: Springer.

Glistau, E., & Coello Machado, N. (2018). Industry 4.0, logistics 4.0 and materials – chances and solutions. *Materials Science Forum, 919*, 307–314.

Govindan, K., & Hasanagic, M. (2018). A systematic review on drivers, barriers, and practices towards circular economy: A supply chain perspective. *International Journal of Production Research, 56*(1–2), 278–311.

Groba, C., Sartal, A., & Bergantiño, G. (2020). Optimization of tuna fishing logistic routes through information sharing policies: A game theory-based approach. *Marine Policy, 113*, 103795.

Guizzi, G., Revetria, R., & Rozhok, A. (2020). Augmented Reality and Virtual Reality. From the Industrial Field to Other Areas. In A. Sartal, D. Carou, & J. Davim (Eds.), *Enabling technologies for the successful deployment of Industry 4.0* (pp. 67–84). Boca Raton, FL: CRC Press.

Holmström, J., Holweg, M., Lawson, B., Pil, F. K., & Wagner, S. M. (2019). The digitalization of operations and supply chain management: Theoretical and methodological implications. *Journal of Operations Management, 65*(8), 728–734.

Hortelano, D., Olivares, T., Ruiz, M., C., G.-H., & López, V. (2017). From sensor networks to internet of things. Bluetooth low energy, a standard for this evolution. *Sensors, 17*, 372.

Horvat, D., Kroll, H., & Jäger, A. (2019). Researching the effects of automation and digitalization on manufacturing companies' productivity in the early stage of Industry 4.0. *Procedia Manufacturing, 39*, 886–893. Elsevier B.V.

Horváth, D., & Szabó, R. (2019). Driving forces and barriers of Industry 4.0: Do multinational and small and medium-sized companies have equal opportunities? *Technological Forecasting and Social Change, 146*, 119–132.

Ivanov, D., Dolgui, A., Das, A., & Sokolov, B. (2019). Digital supply chain twins: Managing the ripple effect, resilience, and disruption risks by data-driven optimization, simulation, and visibility. In D. A. Ivanov D. (Ed.), *Handbook of ripple effects in the supply chain. International series in operations research & management science, vol. 276* (pp. 309–332). Cham: Springer.

Jose, R., & Ramakrishna, S. (2018). Materials 4.0: Materials big data enabled materials discovery. *Applied Materials Today, 10*, 127–132.

Joshi, A., & Gupta, S. (2019). Evaluation of design alternatives of end-of-life products using internet of things. *International Journal of Production Economics, 208*, 281–293.

Kache, F., & Seuring, S. (2017). Challenges and opportunities of digital information at the intersection of big data analytics and supply chain management. *International Journal of Operation and Production Management, 37*(1), 10–36.

Kamble, S., Gunasekaran, A., & Dhone, N. (2020). Industry 4.0 and lean manufacturing practices for sustainable organisational performance in Indian manufacturing companies. *International Journal of Production Research, 58*(5), 1319–1337.

Kamble, S., Gunasekaran, A., & Gawankar, S. (2018). Sustainable Industry 4.0 framework: A systematic literature review identifying the current trends and future perspectives. *Process Safety and Environmental Protection, 117*, 408–425.

Kaswan, M., & Rathi, R. (2020). Investigating the enablers associated with implementation of green lean six sigma in manufacturing sector using best worst method. *Clean Technologies and Environmental Policy, 22,* 865–876.

Kiel, D., Müller, J., Arnold, C., & Voigt, K. (2017). Sustainable industrial value creation: Benefits and challenges of Industry 4.0. *International Journal of Innovation Management,* 21(8), 1740015.

Kirchherr, J., Piscicelli, L., Bour, R., Kostense-Smit, E., Muller, J., Huibrechtse-Truijens, A., & Hekkert, M. (2018). Barriers to the circular economy: Evidence from the European Union (EU). *Ecological Economics, 150,* 264–272.

Koch, P., van Amstel, M., Dębska, P. T., Tetzlaff, A., & Bøgh, S. (2017). A skill-based robot co-worker for industrial maintenance tasks. *Procedia Manufacturing, 11,* 83–90.

Kumar, A. (2018). Methods and materials for smart manufacturing: Additive manufacturing, internet of things, flexible sensors and soft robotics. *Manufacturing Letters,* 15(1), 122–125.

Kusiak, A. (2018). Smart manufacturing. *International Journal of Production Research,* 56(1–2), 508–517.

Lee, I., & Lee, K. (2015). The internet of things (IoT): Applications, investments, and challenges for enterprises. *Business Horizons,* 58(4), 431–440.

Leng, J., Zhang, H., Yan, D., Liu, Q., Chen, X., & Zhang, D. (2019). Digital twin-driven manufacturing cyber-physical system for parallel controlling of smart work-shop. *Journal of Ambient Intelligence Humanized Computing,* 10(3), 1155–1166.

Li, L. (2018). China's manufacturing locus in 2025: With a comparison of "Made-in-China 2025" and "Industry 4.0". *Technological ForecastIng and Social Change, 135*(1), 66–74.

Li, Y., Dai, J., & Cui, L. (2020). The impact of digital technologies on economic and environmental performance in the context of Industry 4.0: A moderated mediation model. *International Journal of Production Economics, 229,* 107777.

Li, Z., Wang, Y., & Wang, K.-S. (2017). Intelligent predictive maintenance for fault diagnosis and prognosis in machine centers: Industry 4.0 scenario. *Advance Manufacturing,* 5(4), 377–387.

Lieder, M., & Rashid, A. (2016). Towards circular economy implementation: A comprehensive review in context of manufacturing industry. *Journal of Cleaner Production, 115,* 36–51.

Lin, K. (2018). User experience-based product design for smart production to empower Industry 4.0 in the glass recycling circular economy. *Computers & Industrial Engineering, 125,* 729–738.

Liu, Y., & Xu, X. (2017). Industry 4.0 and cloud manufacturing: A comparative analysis. *Journal of Manufacturing Science and Engineering, Transactions of the ASME, 139*(3), 034701.

Longo, F., Nicoletti, L., & Padovano, A. (2017). Smart operators in Industry 4.0: A human centered approach to enhance operators' capabilities and competencies within the new smart factory context. *Computers and Industrial Engineering, 1,* 144–159.

Lüdeke-Freund, F., Gold, S., & Bocken, N. (2019). A review and typology of circular economy business model patterns. *Journal of Industrial Ecology,* 23(1), 36–61.

MacArthur, D. E., Zumwinkel, K., & Stuchtey, M. (2015). *Growth within: A circular economy vision for a competitive Europe.* Cowes, UK: Ellen MacArthur Foundation.

MacArthur, D. E., & Waughray, D. (2016). *Intelligent assets. Unlocking the circular economy potential.* Cowes, UK: Ellen MacArthur Foundation.

Majstorovic, V. D., & Stojadinovic, S. M. (2020). Virtualization, Simulation and Cybersecurity – Cloud Manufacturing Issue. In A. Sartal, D. Carou, & J. Davim (Eds.), *Enabling technologies for the successful deployment of Industry 4.0* (pp. 85–104). Boca Raton, FL: CRC Press.

Marconi, M., Michele, G., Mandolini, M., & Favi, C. (2019). Applying data mining technique to disassembly sequence planning: A method to assess effective disassembly time of industrial products. *International Journal of Production Research*, *57*(2), 599–623.

Martín, J. M. M., Martínez, J. M. G., Moreno, V. M., & Rodríguez, A. S. (2019). An analysis of the tourist mobility in the island of Lanzarote: Car rental versus more sustainable transportation alternatives. *Sustainability*, *11*(3), 1–17.

Mattos Nacimento, D., Alencastro, V., Gonçalves, O., Goyannes, R., Garza-Reyes, J., Rocha-Lona, L., & Tortorella, G. (2019). Exploring Industry 4.0 technologies to enable circular economy practices in a manufacturing context. *Journal of Manufacturing Technology Management*, *30*(3), 607–627.

Mattos Nascimento, D. L., Alencastro, V., Quelhas, O. L., Caiado, R. G., Garza-Reyes, J. A., & Rocha-Lona, L. (2018). Exploring Industry 4.0 technologies to enable circular economy practices in a manufacturing context: A business model proposal. *Journal of Manufacturing Technology Management*, *29*(6), 910–936.

McDonough, W., & Braungart, M. (2002). *Cradle to cradle: Remaking the way we make things*. New York (USA): North Point Press.

Moktadir, M. A., Rahman, T., Rahman, M., Ali, S., & Paul, S. (2018). Drivers to sustainable manufacturing practices and circular economy: A perspective of leather industries in Bangladesh. *Journal of Cleaner Production*, *174*, 1366–1380.

Momete, D. C. (2020). A unified framework for assessing the readiness of European Union economies to migrate to a circular modelling. *Science of The Total Environment*, *718*, 137375.

Moreno, M., Carou, D., & Davim, J. (2020). Autonomous Robots and CoBots. Applications in Manufacturing. In A. Sartal, D. Carou, & J. Davim (Eds.), *Enabling technologies for the successful deployment of Industry 4.0* (pp. 47–66). Boca Raton, FL: CRC Press.

Müller, J., Buliga, O., & Voigt, K.-I. (2018). Fortune favors the prepared: How SMEs approach business model innovations in Industry 4.0. *Technological Forecasting and Social Change*, *132*(1), 2–17.

Murray, A., Skene, K., & Haynes, K. (2017). The circular economy: An interdisciplinary exploration of the concept and application in a global context. *Journal of Business Ethics*, *140*(3), 369–380.

Niaki, M., Torabi, S., & Nonino, F. (2019). Why manufacturers adopt additive manufacturing technologies: The role of sustainability. *Journal of Cleaner Production*, *222*(1), 381–392.

Niesen, T., Houy, C., Fettke, P., & Loos, P. (2016). Towards an integrative big data analysis framework for data-driven risk management in Industry 4.0. *2016 49th Hawaii International Conference on IEEE* (pp. 5065–5074). System Sciences (HICSS).

Ormazabal, M., Prieto-Sandoval, V., Puga-Leal, R., & Jaca, C. (2018). Circular economy in Spanish SMEs: Challenges and opportunities. *Journal of Cleaner Production*, *185*, 157–167.

Oztemel, E., & Gursev, S. (2020). Literature review of Industry 4.0 and related technologies. *Journal of Intelligent Manufacturing*, *31*(1), 127–182. Springer.

Pardo, J., Mejías, A., & Sartal, A. (2020). Assessing the importance of biomass-based heating systems for more sustainable buildings: A case study based in Spain. *Energies*, *13*(5), 1025.

Park, S. (2016). Development of innovative strategies for the Korean manufacturing industry by use of the connected smart factory (CSF). *Procedia Computer Science*, *91*(Supplement C), 744–750.

Pheifer, A. G. (2017). Barriers & Enablers to Circular Business Models. *White Paper. Brielle*.

Piran, F., Lacerda, D., Camargo, L., Viero, C., Teixeira, R., & Dresch, A. (2017). Product modularity and its effects on the production process: An analysis in a bus manufacturer. *International Journal of Advance Manufacturing Technology*, *88*(5–8), 2331–2343.

Qi, Q., & Tao, F. (2018). Digital twin and big data towards smart manufacturing and Industry 4.0: 360 degree comparison. *IEEE Access, 6*(1), 3585–3593.

Quenehen, A., Pocachard, J., & Klement, N. (2019). Process optimisation using collaborative robots - comparative case study. *IFAC-PapersOnLine, 52*(13), 60–65.

Quintás, M. A., Martínez-Senra, A. I., & Sartal, A. (2018). The role of SMEs' green business models in the transition to a low-carbon economy: Differences in their design and degree of adoption stemming from business size. *Sustainability, 10*(6), 2109.

Rehman, M., Yaqoob, I., Salah, K., Imran, M., Jayaraman, P., & Perera, C. (2019). The role of big data analytics in industrial internet of things. *Future Generation Computer Systems, 99*, 247–259.

Ren, S., Zhang, Y., Liu, Y., Sakao, T., Huisingh, D., & Almeida, C. (2019). A comprehensive review of big data analytics throughout product lifecycle to support sustainable smart manufacturing: A framework, challenges and future research directions. *Journal of Cleaner Production, 210*, 1343–1365.

Ribeiro, I., Matos, F., Jacinto, C., Salman, H., Cardeal, G., Carvalho, H., … Peças, P. (2020). Framework for life cycle sustainability assessment of additive manufacturing. *Sustainability, 12*(3), 929.

Rönnlund, I., Markus, R., Horn, S., Aho, J., Aho, M., Päällysaho, M., … Pursula, T. (2016). Eco- efficiency indicator framework implemented in the metallurgical industry: Part 1—A comprehensive view and benchmark. *The International Journal of Life Cycle Assessment, 21*(10), 1473–1500.

Salim, H. K., Stewart, R. A., Sahin, O., & Dudley, M. (2019). Drivers, barriers and enablers to end-of-life management of solar photovoltaic and battery energy storage systems: A systematic literature review. *Journal of Cleaner Production, 211*, 537–554.

Sanders, A., Elangeswaran, C., & Wulfsberg, J. P. (2016). Industry 4.0 implies lean manufacturing: Research activities in Industry 4.0 function as enablers for lean manufacturing. *Journal of Industrial Engineering and Management, 9*(3), 811–833.

Sartal, A., & Vázquez, X. H. (2017). Implementing information technologies and operational excellence: Planning, emergence and randomness in the survival of adaptive manufacturing systems. *Journal of Manufacturing Systems, 45*, 1–16.

Sartal, A., Bellas, R., Mejías, A., & García-Collado, A. (2020a). The sustainable manufacturing concept, evolution and opportunities within Industry 4.0: A literature review. *Advances in Mechanical Engineering, 12*(5), 1687814020925232.

Sartal, A., Carou, D., & Davim, J. P. (Eds.) (2020b). *Enabling technologies for the successful deployment of Industry 4.0*. CRC Press, Boca Raton: FL

Sartal, A., Carou, D., Dorado-Vicente, R., & Mandayo, L. (2019). Facing the challenges of the food industry: Might additive manufacturing be the answer? *Proceedings of the Institution of Mechanical Engineers, Part B: Journal of Engineering Manufacture, 233*(8), 1902–1906.

Sartal, A., Martinez-Senra, A., & Cruz-Machado, V. (2018). Are all lean principles equally eco-friendly? A panel data study. *Journal of Cleaner Production, 177*, 362–370.

Sartal, A., Martínez-Senra, A., & García, J. (2017). Balancing offshoring and agility in the apparel industry: Lessons from benetton and inditex. *Fibres and Textiles in Eastern Europe, 25*(2), 16–23.

Sartal, A., Ozcelik, N., & Rodríguez, M. (2020c). Bringing the circular economy closer to small and medium enterprises: Improving water circularity without damaging plant productivity. *Journal of Cleaner Production, 256*, 120363.

Sauerwein, M., Doubrovski, E. L. (2018). Local and recyclable materials for additive manufacturing: 3D printing with mussel shells. *Materials Today Communications, 15*, 214–217.

Schäfers, P., & Walther, A. (2017). Modelling circular material flow and the consequences for SCM and PPC. *Global Journal of Business Research, 11*(2), 91–100.

Schou, C., Andersen, R., Chrysostomou, D., Bøgh, S., & Madsen, O. (2018). Skill-based instruction of collaborative robots in industrial settings. *Robotics and Computer Integrated Manufacturing, 53*(1), 72–80.

Schroeder, A., Ziaee Bigdeli, A., Galera Zarcos, C., & Baines, T. (2019). Capturing the benefits of Industry 4.0: A business network perspective. *Production Planning and Control, 30*(16), 1–17.

Sehnem, S., Jabbour, C., Pereira, S., & de Sousa Jabbour, A. (2019). Improving sustainable supply chains performance through operational excellence: circular economy approach. *Resources. Conservation and Recycling, 149,* 236–248.

Shahbazi, S., Wiktorsson, M., Kurdve, M., Jönsson, C., & Bjelkemyr, M. (2016). Material efficiency in manufacturing: Swedish evidence on potential, barriers and strategies. *Journal of Cleaner Production, 127,* 438–450.

Shankar, K., Kannan, D., & Kumar, P. (2017). Analyzing sustainable manufacturing practices e a case study in Indian context. *Journal of Cleaner Production, 164,* 1332–1343. doi:https://doi.org/10.1016/j.jclepro.2017.05.097.

Song, B., Yeo, Z., Kohls, P., & Herrmann, C. (2017). Industrial symbiosis: Exploring big-data approach for waste stream discovery. *Procedia CIRP, 61,* 353–358.

Srivastava, S. K. (2007). Green supply-chain management: A state-of-the-art literature review. *International Journal of Management Reviews, 9*(1), 53–80.

Stahel, W. R., & Reday-Mulvey, G. (1981). *Jobs for tomorrow: The potential for substituting manpower for energy.* New York (USA): Vantage Press.

Stock, T., & Seliger, G. (2016). Opportunities of Sustainable Manufacturing in Industry 4.0. *Procedia CIRP, 40,* 536–541.

Stone, C., Neely, A., & Lengnick-Hall, M. (2018). Human Resource Management in the Digital Age: Big Data, HR Analytics and Artificial Intelligence. In P. Melo, C. Machado (Ed.), *Management and technological challenges in the digital age* (pp. 13–42). CRC Press. Boca Raton: FL

Su, B., Heshmati, A., Geng, Y., & Yu, X. (2013). A review of the circular economy in China: Moving from rhetoric to implementation. *Journal of Cleaner Production, 42,* 215–227.

Sun, E., Wang, F., & Wang, H. (2019). The safety training system based on AR cloud in metallurgical enterprises. *Journal of Industrial and Intelligent Information, 7*(2), 37–41.

Tan, Y., Shen, L., & Yao, H. (2011). Sustainable construction practice and contractors' competitiveness: A preliminary study. *Habitat Int., 35,* 225–230. doi:https://doi.org/10.1016/j.habitatint.2010.09.008.

Tao, F., Cheng, J., Qi, Q., Zhang, M., Zhang, H., & Sui, F. (2018). Digital twin-driven product design, manufacturing and service with big data. *International Journal of Advance Manufacturing Technology, 94*(9–12), 3563–3576.

Thoben, K., Wiesner, S., & Wuest, T. (2017). "Industrie 4.0" and smart manufacturing – a review of research issues and application examples. *International Journal of Automation Technology, 11*(1), 4–16.

Tiwari, K., & Khan, M. (2020). Sustainability accounting and reporting in the Industry 4.0. *Journal of Cleaner Production, 258,* 120783.

Torn, I., & Vaneker, T. (2019). Mass personalization with Industry 4.0 by SMEs: A concept for collaborative networks. *Procedia Manufacturing, 28,* 135–141. Elsevier B.V.

Tortorella, G., & Fettermann, D. (2018). Implementation of Industry 4.0 and lean production in Brazilian manufacturing companies. *International Journal of Production Research, 56*(8), 2975–2987.

Ülkü, M., & Hsuan, J. (2017). Towards sustainable consumption and production: Competitive pricing of modular products for green consumers. *Journal of Cleaner Production, 142*(1), 4230–4242.

Usuga Cadavid, J., Lamouri, S., Grabot, B., Pellerin, R., & Fortin, A. (2020). Machine learning applied in production planning and control: A state-of-the-art in the era of Industry 4.0. *Journal of Intelligent Manufacturing, Springer. 31*, 1531–1558

van Schaik, A., & Reuter, M. A. (2016). Recycling indices visualizing the performance of the circular economy. *World of Metallurgy-ERZMETALL, 69*(4), 201–216.

van Wijk, A. D., & van Wijk, I. (2015). *3D printing with biomaterials: Towards a sustainable and circular economy.* Amsterdam (Netherrland): IOS Press BV.

Vanner, R., Bicket, S., Withana, P., Ten Brink, P., Razzini, E., van Dijl, E., … Hudson, C. (2014). Scoping Study to Identify Potential Circular Economy Actions, Priority Sectors, Material Flows & Value Chains (DG Environment's Framework Contract for Economic Analysis ENV.F.1/FRA/2010/0044 No. Final Report). Policy Studies Institute (PSI), Institute for European Environmental Policy (IEEP), BIO and Ecologic Institute.

Voet, V. S., Strating, T., Schnelting, G. H., Dijkstra, P., Tietema, M., Xu, J., & … & Folkersma, R. (2018). Biobased acrylate photocurable resin formulation for stereolithography 3D printing. *ACS omega, 3*(2), 1403–1408.

Wagner, T., Herrmann, C., & Thiede, S. (2017). Industry 4.0 impacts on lean production systems. *Procedia CIRP, 63*, 125–131.

Yao, X., Zhou, J., Lin, Y., Li, Y., Yu, H., & Liu, Y. (2019). Smart manufacturing based on cyber-physical systems and beyond. *Journal of Intelligent Manufacturing, 30*(8), 2805–2817.

Yin, Y., Stecke, K., & Li, D. (2018). The evolution of production systems from industry 2.0 through Industry 4.0. *International Journal of Production Research, 56*(1–2), 848–861.

Zheng, P., Sang, Z., Zhong, R., Liu, Y., Liu, C., Mubarok, K., … Xu, X. (2018). Smart manufacturing systems for Industry 4.0: Conceptual framework, scenarios, and future perspectives. *Frontiers in Mechanical Engineering, 13*(2), 137–150.

3 Reporting for New Business Models
The Challenge to Support the Circular Economy

Elaine Conway
Derby Business School, University of Derby,
Derby, UK

CONTENTS

3.1 INTRODUCTION

Over the last few years, businesses have demonstrated an increasing awareness of the need to address environmental issues within the scope of their activities (Ünal *et al.*, 2019). This has manifested itself in a range of activities within businesses, such as applying for formal International Standards Organisation (ISO) or European Eco-Management and Audit Scheme (EMAS) accreditations to implement environmental management (Centobelli *et al.*, 2020) or supply chain audits to drive out waste. Firms are increasingly realising that the raw materials they input into their processes need as much attention as the production process itself, given the increasing scarcity and cost of many of these raw materials (Rosa *et al.*, 2019). This is necessitating businesses to review the source and application of their raw materials and to focus on recycling more materials back into the production system. Some of this is being mandated by laws such as the Waste Electrical and Electronic Equipment Regulations 2018 (UK Government, 2018), which mandates the 'take-back' of used consumer goods.

To support the measurement and recording of these changes, there have been developments in accounting, including material flow accounting and analysis

(Bebbington *et al.*, 2007; Bringezu *et al.*, 2003; Gray & Bebbington, 2001; Larrinaga-González *et al.*, 2001). However, these advances in measurement have been largely focused on the management accounting side of the business, providing richer data on the breadth of organisational impacts on the environment and society.

Thus far, that same support has not been matched in the financial reporting sphere at individual firm level. Whilst the International Accounting Standards Board (IASB) and its US counterpart the Financial Accounting Standards Board (FASB) have issued new financial reporting standards during this time, they have not addressed the changing perspectives of business and business models in the same way. Clearly, the time it takes for these institutions to gain worldwide approval for new standards means that they react much more slowly to changes in the business environment. Nonetheless, this is an area that needs to be addressed as new business models, such as those based on the circular economy (CE) gain greater traction in the business world as their impacts on the firm's financial reports can be substantial.

This chapter examines the challenges of financial reporting to support the CE. There is a plethora of academic work on Social and Environmental Accounting (SEA); however, much of this work is largely focused on the output (production) or waste flow side of the production process (e.g. greenhouse gas emissions or carbon dioxide equivalents). There is very little acknowledgement of the effect moving to CE might have on the financial reports of the business and how firms might mitigate these effects to maintain continued shareholder and stakeholder support.

3.2 EVOLVING BUSINESS MODELS

The change in perception about having to design products for recovery and recycling after use is bringing about changes in business models (Frishammar & Parida, 2019; Geissdoerfer *et al.*, 2018). Firms are starting to engage with issues such as product longevity and raw material re-use, either within their own facilities or through third party contractors (Ünal *et al.*, 2019). This causes a change in the way in which firms recognise assets, liabilities, costs and revenue in their financial reports, because the relationship between the manufacturer/seller and consumer of the goods is changing. Where once it was a one-off transaction with some limited aftersales or warranty provision, it is now becoming a longer term, repeat relationship with greater requirements for servicing and return (Rosa *et al.*, 2019).

International Financial Reporting Standards (IFRS) and US Generally Accepted Accounting Principles (US GAAP) currently support a largely traditional business model, at least where manufacturing is concerned. This follows the linear approach (as indicated in Figure 3.1), where raw materials are extracted, sold to parts manufacturers, then sold on to final product manufacturers, from where the final goods are shipped and sold either directly to the final consumer or via a distributor or service company (Tukker, 2015).

This business model is very simple and does not address the waste streams at any part of the chain beyond some regard for scrap valuation during the production process. It does not seek to evaluate any form of recycling or re-use of material back into the process. At worst, it can be seen as a method of producing waste from natural resources through the process of manufacturing (Murray *et al.*, 2017). This

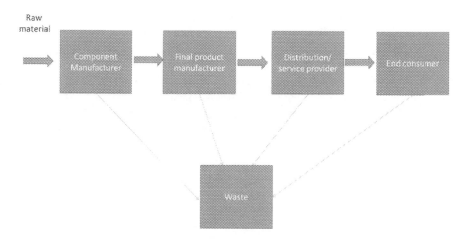

FIGURE 3.1   Model of a simple linear production system.

process, if left unchecked, will cause the environment to deteriorate by depleting valuable natural resources and then leaving behind waste and pollution (Boulding, 1966; Ellen Macarthur Foundation, 2012b; Rosa *et al.*, 2019; Ünal *et al.*, 2019). Many current attempts by businesses to manage their impacts have been focused on 'end-of-pipe' solutions (Larrinaga *et al.*, 2002). Whilst these mitigate the immediate impact of operations by reducing pollution or waste, they do not address the scarcity of materials or the drive to more sustainable energy sources. Longer term, this limited reductionist viewpoint will threaten continued economic prosperity, social wellbeing and human health (Ellen Macarthur Foundation, 2012b; Gray, 2002; Rosa *et al.*, 2019).

Whilst there has been increased reporting on these environmental measures and greater attention to social issues in Corporate Social Responsibility (CSR) reports over recent years, this can often equate to 'business as usual' beneath the surface (Frishammar & Parida, 2019; Gray, 2002, 2006), with no real long-term business restructuring to address the concerns of dwindling natural resources and an increasing global population. Indeed, as Gray (2002) postulates, accountants, with their narrow reductionist view on producing standards that focus on being consistent and only evaluating business in financial terms, find it challenging to get to grips with the concept of sustainability and systems thinking. Even this recent broadening out of the reporting of social and environmental aspects of businesses tends to be focused on the material impacts these issues might bring to the economic fortunes of the firm (Larrinaga *et al.*, 2002). Hence this further inhibits wider consideration of the interconnectedness and interdependencies business has with its environment (Larrinaga-González *et al.*, 2001), which are not so readily translated into financial impacts.

Since most investors in the Western world are still intrinsically motivated by financial return, despite the growth in sustainable business investing (Larrinaga *et al.*, 2002), whilst materials remain relatively cheap and costs to recycle relatively high, there is little incentive for firms to invest beyond some legal minimum recycling requirement (Anderson, 2007). Clearly as those raw materials become scarcer and hence more expensive, and society recognizes the finite nature of those resources, that standpoint will change, whether through a policy-driven environmental tax 'push' or the fear of losing access to key resources (Gray, 2002; Larrinaga-González *et al.*, 2001; Rosa *et al.*, 2019). This change is undoubtedly causing firms to assess how they could re-use more materials back into their production process.

Whilst the concept of circularity and re-use of materials is not new (Lieder & Rashid, 2016), the CE as an emerging business model is only relatively recently gaining traction in the business world (Murray *et al.*, 2017). Faced with increasing shortages of raw materials, particularly rare earth minerals, manufacturers are starting to realise that the precious raw materials incorporated in previously manufactured goods are a source of raw materials in future goods. Indeed, the Chinese government, having acknowledged their economic boom has been supported largely at the expense of the environment, have adopted the CE as the basis for their national policy since 2002 (Geng *et al.*, 2012; Lieder & Rashid, 2016; Murray *et al.*, 2017).

3.3   WHAT IS THE CIRCULAR ECONOMY?

The concept of the CE centres around the idea of resource cycling (Geng & Doberstein, 2008; Ghisellini *et al.*, 2016; Murray *et al.*, 2017). Engaging a combination of increasing the longevity of products through better product design, better materials usage, better manufacturing and maintenance and reusing materials, the rate of depletion of basic raw materials is reduced (Tukker, 2015; Tukker *et al.*, 2006; Ünal *et al.*, 2019). Waste from other products becomes 'food' for other processes or products through the concepts of Reduce, Reuse and Recycle (the three R's) (Ellen Macarthur Foundation, 2012b; Lieder & Rashid, 2016; Murray *et al.*, 2017). Hu *et al.* (2011) have extended this concept to include a fourth 'R', 'Recover', to incorporate once-used materials back into a usable state in the production chain. Other authors have even extended this to six R's or principles, reuse, reduce, recycle, redesign, remanufacture and repair (Lüdeke-Freund *et al.*, 2019).

Overall, however, the emphasis is on using less material in the production cycle, making products more durable and using any waste that is produced back into the process (or circle) again (Ellen Macarthur Foundation, 2012b). The key to this is designing circularity of materials and products by ensuring that the basic raw materials do not get contaminated and are easy to retrieve at the end of the cycle, thus enabling them to be readily incorporated into the cycle again without any degradation of quality (Ellen Macarthur Foundation, 2012b; Ghisellini *et al.*, 2016). This process is modelled on biomimicry of the natural ecosystem, which naturally recycles waste back into the natural lifecycle (Geng *et al.*, 2008).

An example of a simple CE model is depicted in Figure 3.2.

The CE approach does not exist without its critics. For example, with regards to the broad agenda of sustainable business, CE remains silent on the matter of social responsibility (Murray *et al.*, 2017). Whilst tacitly it is present from an overall standpoint for making the environment better for all, more active social responsibility initiatives are not inherent to CE, as it comes from an industrial ecology background (Anderson, 2007; Ghisellini *et al.*, 2016; Murray *et al.*, 2017; Ünal *et al.*, 2019). Therefore, it is not clear how CE might tackle social inequality issues or facilitate

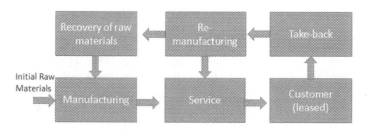

FIGURE 3.2   Model of a simple circular economy model.

greater social opportunity. That said, Frishammar and Parida (2019) felt that there is social value in changing customer behaviour from needing to own assets, to sharing their use. Rosa *et al.* (2019) also advocated some social benefits of a CE approach: the development of innovative skills and knowledge, the improvement of health and safety in workplaces, and increased employee motivation through an enhanced corporate reputation.

Equally, there is criticism that by designing more durable products that they may not degrade as well at the final end of their life or may become too complex to strip down to retrieve useful raw materials from the end product (Murray *et al.*, 2017). Further, whilst the CE approach is focused on recycling materials, there is nonetheless a limit; eventually there will be some waste that cannot be utilised by the system (Anderson, 2007; Pedersen *et al.*, 2019), but then the emphasis should be on ensuring that that resultant waste has as minimal impact on the environment as possible.

Currently, most attempts at the CE have focused on creating new products from waste, rather than the original manufacturers taking back their own products (Ghisellini *et al.*, 2016), which is the ideal manifestation of the CE. This approach requires fewer resources, energy and labour in comparison to the manufacture of goods using virgin materials (Ghisellini *et al.*, 2016; Pedersen *et al.*, 2019). The barrier to this at the moment is that the waste collection systems are largely controlled by waste management companies, rather than the manufacturers themselves (Singh & Ordoñez, 2016).

Under a CE, the idea is that for durable goods, such as washing machines, cars and even mobile phones, the business model will change. Manufacturers will lease these products, not sell them outright as the majority do today (Tukker, 2015; Ünal *et al.*, 2019). They will provide a full service and maintenance contract over the life of the leased asset (further encouraging manufacturers to design quality, longevity and ease of servicing into their products). Some business models may even contract with customers on the basis of pay-per-use, with the internet of things supporting intelligent billing (Centobelli *et al.*, 2020; Ünal *et al.*, 2019). At the end of the lease period, manufacturers will take back products and re-manufacture/re-condition them and lease them out again, hence extending the life of the assets substantially. To underpin this, the expectation is that manufacturers will be encouraged to design in quality and longevity, together with the ability to service, maintain and recover key components within their products (Ellen Macarthur Foundation, 2012a, 2012c, 2013; Frishammar & Parida, 2019; Rudnicka, 2018).

This increased number of loops back into the business from recycling, remanufacturing and re-leasing will undoubtedly complicate the business model compared with the traditional linear model, as shown in Figure 3.1. Depending on the products concerned, there could be a substantially longer material flow that lasts for years, rather than potentially a few months under current systems, as depicted in Figure 3.2. It will require manufacturing firms to create service departments (or outsource them) to support long-term customer leasing contracts, tracking of assets at all points of their lives, arranging service, maintenance and recovery of assets, a re-manufacturing service for used assets, and a sales and marketing team to promote the use of re-manufactured products and the concept of rental, rather than ownership of certain assets.

3.4 TRANSITIONING TO THE CE

Business models are ways of communicating the sources of value that firms create (Pedersen *et al.*, 2019; Teece, 2010). There are various reasons for firms wanting to transition away from a linear open manufacturing model to a more sustainable business model, such as a closed loop one based on the CE. These can include realisation of the reliance on ever-decreasing natural resources, pressure from supply chain partners (Centobelli *et al.*, 2020; Ünal *et al.*, 2019), volatile commodity prices, competitive threats, legislation or a desire on the part of management to make the change (Frishammar & Parida, 2019).

The main areas that business models should cover include value creation, value transfer and value capture (Centobelli *et al.*, 2020; Geissdoerfer *et al.*, 2018; Urbinati *et al.*, 2017). In the standard linear model, value creation is traditionally focused on the one-time sale of a product to a customer. However, the focus in a CE business model is to retain that value creation within the system for longer through increased durability, use and re-use (Centobelli *et al.*, 2020; Geissdoerfer *et al.*, 2018). It will require firms to deliberately narrow, slow and close material and energy flows, use renewable energy sources and materials in order to reduce waste and maintain materials at the highest possible value for the longest time possible in the supply chain (Pieroni *et al.*, 2019).

Value transfer relates to the customer-centric elements of the product offer, such as customer segmentation and customer relationships and requires a strong focus on product design to address customer expectations but also durability, environmental impact and cost over of the whole life cycle of the product (Geissdoerfer *et al.*, 2018; Pedersen *et al.*, 2019). Value capture focuses on obtaining additional revenue streams, either from cost reductions, product or process restructures but also from potential extensions to the market (Centobelli *et al.*, 2020; Frishammar & Parida, 2019).

The aim of the CE is to maximise the lifecycle of a given product throughout its supply chain (Centobelli *et al.*, 2020; Frishammar & Parida, 2019). CE encourages firms to review how they create, deliver and capture value through their products and processes (Ünal *et al.*, 2019). CE value creation centres on maintaining products and the use of upcycling and recycling waste, reducing material content and production on demand rather than constant over-production (Centobelli *et al.*, 2020; Pieroni *et al.*, 2019) and then providing customers with a service, rather than a fixed product.

Transitioning to CE requires a long-term view and closer working with wider stakeholder networks (Geissdoerfer *et al.*, 2018; Pedersen *et al.*, 2019). Adopting a product–service system (PSS) approach is perceived as key to increase resource efficiency throughout the lifecycle of the product from material extraction through to end-of-life treatment (Frishammar & Parida, 2019; Ünal *et al.*, 2019). PSS-based business models are predicated on a rental/leasing or pay-per-use model of value creation rather than the exchange of ownership of the product (Frishammar & Parida, 2019; Tukker, 2015; Ünal *et al.*, 2019). Hence this requires a particular change in focus of what is being offered to the customer, how it is to be delivered (by paying per hour for a service or a deliverable, rather than a physical product) and evaluating whether the revenue sources from these services are sustainable (Frishammar & Parida, 2019; Tukker, 2015). This might well involve other suppliers of services to support the business offering, which will require appropriate management to maintain the firm's reputation.

The transition to CE presents numerous other challenges; it may require substantial cultural change both internally and externally to the firm. Traditional product-based revenue streams may decline in importance and newer service-orientated approaches will assume a greater importance in the firm (Frishammar & Parida, 2019; Pedersen *et al.*, 2019). There may also be concerns that remanufactured goods could cannibalise the sales of new products (Pedersen *et al.*, 2019). This may also require changes to incentives and reward systems as sales staff may not be so keen to sell re-manufactured or re-conditioned goods (Hopkinson *et al.*, 2018). Equally, sales staff will need to manage a higher number and a different range of interactions with their customers as their relationships may change from being largely one-off sales interactions to more long-term service and maintenance contacts through leasing and rental contracts (Urbinati *et al.*, 2017). Hence corporate reward systems may have to evolve to address this changing customer support network.

From a customer perspective, some cultures (e.g. France and Germany) are more accepting of both returning products to be recycled and in buying recycled goods in comparison to others (e.g. the UK). They may also have well-developed infrastructures to allow customers to return goods more easily than others (Pedersen *et al.*, 2019). To support the move to a CE business model, governments may have to incentivise consumers to use appropriate waste separation and collection systems to minimise the cost of recycling but also to improve the flow of materials back into the system (Rudnicka, 2018).

From a sales perspective, branding and pricing remanufactured or re-conditioned products can be a challenge (Hopkinson *et al.*, 2018). Having encouraged customers to believe that new goods are the best for many years, marketers and sales staff may have difficulty in convincing customers that remanufactured goods are good quality or economically appealing (Ünal *et al.*, 2019; Urbinati *et al.*, 2017). Hence there may not be a sustainable market for such goods, at least in the short term until attitudes change (Lahane *et al.*, 2020). Without strong assurances about quality, remanufactured goods may damage corporate reputation (Pedersen *et al.*, 2019).

Ricoh, the Japanese printer and imaging manufacturer, encountered these issues when selling its range of remanufactured copier machines alongside their new ones (Hopkinson *et al.*, 2018). However, they were able to manage quality concerns by guaranteeing their remanufactured machines against an international quality standard, BS8887-220. This states that remanufactured goods must retain their original specification, but that the packaging must reflect that it may contain recycled or reused parts (Hopkinson *et al.*, 2018). The company was also able to direct its remanufactured copier machines to more price sensitive customers (local government departments) and markets (such as emerging countries), reducing possible cannibalism of its primary new product markets.

Another perceived impediment for encouraging customers to accept renting or leasing assets rather than purchasing them is the concern that by renting goods rather than buying, customers may not take advantage of any technological changes that occur, such as efficiency gains, improved materials, etc., when they are locked into a contract (Ellen Macarthur Foundation, 2012a). Indeed, once the responsibility for taking back goods rests with manufacturers, they may even be discouraged from making significant product improvements when they are faced with the necessity to take back goods that may be obsolete. Firms will need to consider this in their product development, particularly when new products may be developed that are more efficient than the remanufactured ones (Hopkinson *et al.*, 2018).

Another challenge of transitioning to a CE where firms have both new and remanufactured goods in their portfolio is managing production. When all sales are based on new goods, demand is easier to manage; however, with both primary manufacturing and remanufacturing, managing supply and demand of both product ranges can make production and sales target-setting challenging, potentially leading to excess inventory or stock-outs (Hopkinson *et al.*, 2018).

There may also be upfront investments required to repair and remanufacture goods, or for training and recruitment to counter a lack of skills to re-manufacture to the appropriate standard (Ghisellini *et al.*, 2016; Pedersen *et al.*, 2019; Pieroni *et al.*, 2019). This investment might also extend infrastructure to support a shift to renewable energy, as the focus of the CE is not just preserving energy and fuel or becoming more efficient but using sustainable energy sources (Frishammar & Parida, 2019).

Other challenges in the adoption of the CE include the paucity of environmental laws or other regulations in some jurisdictions (Pedersen *et al.*, 2019), a lack of government support and disinterest by top management (Lahane *et al.*, 2020).

3.5   THE FINANCIAL REPORTING CHALLENGE FROM THE CE

At the firm level, the traditional linear route, as depicted in Figure 3.1, is adequately supported by current financial standards and is understood by auditors and investors. Currently, goods for sale are valued under International Accounting Standard (IAS) 2 *Inventory* (IASB, 2016a), capturing value from initial purchase of raw material, labour and production costs throughout the manufacturing process. Once title to the goods has passed, a simple sales transaction is recorded, either on credit or for cash. Generally, depending precisely on whether the firm sells business-to-business or business-to-consumer, there is little further interaction between customer and the firm beyond some occasional aftersales service or warranty claim.

However, as noted earlier, the move to a CE at a business level will undoubtedly create a more complicated business model than at present (Ghisellini *et al.*, 2016). If the ultimate business model for standard consumer goods, such as fridges and washing machines, is for firms to lease, rather than sell their products, and provide a comprehensive service and maintenance contract over the lease period (Centobelli *et al.*, 2020), then the financial reporting implications of that change will be substantial. Manufacturers will have to address the need to design more durable and serviceable products (Centobelli *et al.*, 2020), which have the ability to be re-manufactured and re-used, which may cause initial production cost to rise, which will impact on short-term profitability.

Additionally, firms will need to factor in servicing and maintenance provision whilst assets are under lease and to create take-back systems to retrieve products at the end of their lease period or if faulty under warranty. Again, this may require upfront investment in infrastructure and skills to enable this, and could negatively impact financial returns, with higher costs and higher asset bases potentially eroding profitability and shareholder returns. Whilst many manufacturing firms may not choose to create these service or take-back systems themselves and outsource them to other suppliers, nonetheless, they will be inevitably taking back previously used raw materials at various points in their supply chain and dealing with inventory movement and valuation issues that they do not engage with at present.

From a financial accounting standpoint, this will affect the firm in significant ways. Moving away from a simple sales model to a leasing model will have the effect of calculating revenue recognition from a simple transaction one-off transaction to a multi-year model with a series of lease payments being made (Frishammar & Parida, 2019), subject to the appropriate calculation of finance charges. This will potentially change the recognition of associated costs in order to match costs with the revenue streams they are associated with. Equally, moving to a more service-based model may make revenue streams less secure with an increase in potential defaults on customer lease payments.

These multiple avenues of revenue will require management and a change from a product to a service-based business model (Frishammar & Parida, 2019). They will also need to be sufficient for the firm to be sustainable. The firm will need to provide guarantees of service provision or availability of the use of a product instead of ownership ((Frishammar & Parida, 2019). The revenue implications are that there may be a need for initial cash flows to exit the firm to invest in the infrastructure required to support the CE, prior to cash inflows. These inflows will be for lower monthly amounts, spread over a longer time frame (the lease term) than from the one-off sales model. This may well reduce both project appraisal and payback results in contrast to the conventional one-off sales model. It is important therefore for investors to understand the implications of this in terms of expected shareholder returns, which could be lower in the short term, but more sustainable in the long term.

The other challenge is the valuation of products during their lives. Goods for sale are valued under IAS 2 *Inventory* as previously mentioned; however, under a leasing model, where the assets will be potentially leased out for several years, they will now need to be capitalised, increasing overall asset values on the statement of financial position (SOFP) and making them subject to standards such as

IFRS 16 *Leases* (IASB, 2016b). This raises the question of the valuation of assets which are continually recycled and remanufactured; firms will need to evaluate the costs of this rework and remanufacture (which may often vary substantially through the life of the asset) on a product by product basis – which could be very challenging for auditors and investors to understand as the manufacturing process may well be quite complicated. There is little guidance on this kind of constant renewal of traded assets under current IAS.

Associated with the move of firms from a traditional sales business model to a leasing business model could be increased deferred income on firm balance sheets (for any potential customer deposits), the necessity to split out service costs within leasing contracts and the increased risk of bad debts in their financial statements, which would need to be fully explained to investors. This may make effective analysis and comparisons of firms difficult, since these balances will undoubtedly affect ratios such as return on assets and return on equity. It may also be imprudent to use this group of leased-out assets as the basis for securing debt finance since there would be uncertainty about their value, particularly if they have been re-conditioned several times. The assets may not be readily recoverable from customers' premises, as Ricoh found with approximately 40% of its printers not being recovered at the end of the contract term (Hopkinson *et al.*, 2018).

Equally, firms may be challenged in setting different prices for 'new' products made from completely new materials versus a reconditioned product (Hopkinson *et al.*, 2018; Pedersen *et al.*, 2019). Currently, if the current economic life of a consumer good such as a washing machine is 10 years (Stahel, 1991), under the linear-manufacturing-sales business model, the price is usually set with reference to the costs of making the product. There is usually little or no consideration of the good's scrap value since the customer would tend to dispose of it through waste collection systems or take-back schemes, which dispose of the waste often with minimal recycling. However, if that washing machine were then taken back by the manufacturer and re-conditioned with new replacement parts such as motors and timers (which are the parts most likely to wear) (Ellen Macarthur Foundation, 2012a; Stahel, 1991), firms will need to evaluate the cost of doing this and work out how to set prices for the re-conditioned goods. Logically, if the parts which wear are replaced each time, there is no reason why that same basic good could not last another ten years. This was the approach adopted by Ricoh with its remanufactured copiers (Hopkinson *et al.*, 2018). If supported by the same warranty as the original machine, firms will be faced with dilemma of considering different pricing structures for each subsequent time an asset is leased out. This could substantially complicate the pricing structure and hence revenue recognition of income on assets which are leased out multiple times.

In some industries, for example, wedding clothing hire (where goods have a much shorter life span than consumer goods), there is no price differential for hiring brand new clothing versus clothing which has been hired out previously – in most cases, the customer is not aware whether they have new or previously hired clothing. However, it is unlikely that this would be universally acceptable for consumers for all goods. Firms will need to propose value to customers based on functionality, rather than ownership (Centobelli *et al.*, 2020). It may require a significant re-think on the part of manufacturer and consumer that it is a service which is essentially being sold (e.g. a number of washes or a washing service) rather than a new washing machine. It is this shift of business model which financial accountants need to consider.

From a commercial standpoint, moving from a sales to a more service-based business model will create the need for additional business activities throughout the supply chain (Geissdoerfer *et al.*, 2018; Ünal *et al.*, 2019; Urbinati *et al.*, 2017). These will include the servicing and maintenance of assets and the collection of assets from customer premises at the end of the lease. It will also require legal teams to draw up customer contracts and to ensure there is a clarity over asset insurance, and accountants to monitor customer accounts and manage issues of non-payments or defaults on lease payments. Assets will need to be tracked throughout their lifetime, and auditors may be faced with the reality that large amounts of inventory are not readily verifiable in customer homes.

Equally, firms may have to invest in substantial information technology to monitor asset movements. Advances in technology may well support this diffuse asset tracking (Frishammar & Parida, 2019), with innovations such as radio frequency identification (RFID) (Hedgepeth, 2006), blockchain technology (Andersen, 2016; Ellen MacArthur Foundation, 2016; Percy, 2016) and the Internet of Things (Centobelli *et al.*, 2020) allowing firms to track and audit asset movements and interventions to assets (lease, re-lease, remanufacture, service) during their lifetimes. For example, charge-per-use technology, such as embedding computer chips into tyres to track kilometres driven, enabled Michelin to charge customers based on exact tyre usage (Frishammar & Parida, 2019). Technology will also help firms to maintain communication with their supply chain, which is of particular importance to establish and maintain service components within leased/hired business models if firms do not carry out the whole process themselves (Frishammar & Parida, 2019). Customer management software will be required to allow firms to track customer accounts and payments, block access to assets where payments are outstanding, hence reducing the need to chase overdue payments.

Doubtless, many firms will subcontract some of these processes to other firms (Frishammar & Parida, 2019), but nonetheless, there will still need to be some recognition of the need to review recycled asset and material valuations throughout the supply chain to ensure that underlying business values are realistic and continue to show a true and fair view of the business.

Managing relationships with shareholders is traditionally the remit of accountants. Many firms will adopt mixed business models with multiple revenue streams and sources of value creation, such as traditional product-focused sales but also service or solution-based revenue streams from activities such as leasing, pay-per-use (Frishammar & Parida, 2019). Understanding these multiple sources of value creation and their importance to the economic fortunes of the firm will require considerable communication to a wider range of both shareholders and stakeholders (Geissdoerfer *et al.*, 2018; Pedersen *et al.*, 2019), for whom such approaches are unfamiliar. For example, investors unsure of the new business model may need more convincing of the need for upfront investment in machines and training to acquire the necessary infrastructure to remanufacture: this may cause initial negative cash flows prior to establishing a sustainable customer base for pay-per-use or leasing income (Pieroni *et al.*, 2019). In contrast to a one-time sales model, these differing cash flows and revenue recognition patterns with a pay-per-use or leasing model may cause misunderstandings by investors.

However, research suggests that firms and shareholders should see longer term economic benefits from a CE business model. This may include an overall reduction in costs from raw materials right through to transport of finished goods, increased resource efficiency by using renewable resources and eliminating waste, and reduced product and process complexity (Rosa *et al.*, 2019). It should also reduce risk through improved reputation, business resilience (through lower risk of operational interruption due to resource scarcity), and improved compliance with environmental regulation through reduced environmental impacts (Rosa *et al.*, 2019). It may also open up new revenue streams and markets, improve competitive advantage in terms of products and services offered and enhance supply chain collaborations (Hopkinson *et al.*, 2018; Rosa *et al.*, 2019).

3.6   INTEGRATED REPORTING TO SUPPORT THE CE

Given the impact that moving to a CE business model could have on the financial reporting of a firm, finance directors and other senior management need to communicate clearly with shareholders and other stakeholders about the value creation within their business model to gain and maintain their continued support (IIRC, 2011). Traditional corporate reporting comprises the financial statements and a series of supplementary reports, such as CSR reports. They are often very lengthy, and their focus remains on financial data. Approaches such as integrated reporting (IR) may well support CE better than traditional methods of financial reporting as there is greater scope for the firm to explain their sources of value creation to not only shareholders, but other stakeholders, over a longer time frame than just the past financial reporting year (Conway *et al.*, 2020).

An integrated report is defined as 'a concise communication about how an organisation's strategy, governance, performance and prospects, in the context of its external environment, lead to the creation of value in the short, medium and long term' (IIRC, 2019b). It is this longer-term view that would support the CE, given the much longer timeframe over which the CE is aimed. IR is also specifically predicated on identifying the value creation of the firm, which again supports the CE ethos. The idea is that in their integrated report, firms explain how value creation is maintained across six capitals, financial, manufactured, intellectual, human, social and relationship, and natural (IIRC, 2019a), hence the focus is not exclusively financial, unlike traditional financial reporting. Examples of these capitals are shown in Table 3.1.

These capitals are recognised as inputs to the business model of the firm, which are then transformed through the firm's business activities, guided by its mission and vision, into outcomes, which should equally be measured across each of the six capitals (Conway, 2019; Freeman *et al.*, 2004). The underpinning ethos of IR is that value should be created or at least preserved (not diminished) across the six capitals, although there can be transfer between them (IIRC, 2019a). For example, financial capital can be transferred to manufactured capital when a new machine is purchased.

TABLE 3.1
The Six Capitals of Integrated Reporting (Adapted from EY (2016))

Capital	Encompasses
Natural	Natural resources, such as water, fossil fuels, solar energy, crops and carbon sinks. These cannot be readily replaced and are vital to the normal functioning of the planet.
Social and relationship	The relationships and resources between the organisation and its stakeholders, including shareholders, customers, suppliers, government and community.
Intellectual	Intangible assets linked to the brand and reputation of the organisation, and includes patents, copyright, organisational systems, policies and procedures.
Human	The skills and knowledge of the staff of the organisation, including their motivation and commitment which affect their ability to leverage their skills.
Financial	Financial means to maintain and expand the organisation, such as funds through financing (equity or loans) or internally generated funds.
Manufactured	The physical infrastructure of the organisation, together with its associated technology, tools and equipment.

Equally appropriate to the increased networks created by CE is the concept of integrated thinking which underpins IR. This examines the relationships which are essential to the success of the firm and how they cooperate to achieve the long term aims of the business (Nylander, 2015). It demonstrates the interdependencies between non-financial and financial issues and how they impact financial performance. Once shareholders have an understanding of how the firm creates value, across both financial and non-financial metrics, then they are more likely to adopt a longer-term view in their investing behaviour (Kwan in Nylander (2015)).

Whilst IR appears a good fit with the CE business model, it is far from universally adopted. Despite strong support from the International Integrated Reporting Council (IIRC) and IR being made mandatory for listed companies in the South African stock market, its voluntary nature has impeded its adoption beyond a few thousand companies worldwide (Rinaldi et al., 2018). This is because of the costs to implement, concerns about lack of universal standards which could impede assurance to investors (Adams & Simnett, 2011; Eccles et al., 2012).

These issues notwithstanding, many investors are generally favourable to IR (Rinaldi et al., 2018), as it is regarded as more concise and clear way for the firm to tell its story of value creation and to clarify its business model (Adams & Frost, 2008). Hence given the impacts to the firm's financial performance as discussed above and the more complex business models that may emerge from the CE, IR could mitigate misunderstandings with shareholders and wider stakeholders to engage them with their new ways of working.

3.7 CONCLUSION

The aim of this chapter is to encourage debate on the issues of financial reporting stemming from the impacts that new CE business models will undoubtedly have on the financial performance of the firm in the future. Whilst much work has been accomplished in measuring, monitoring and managing many environmental and social accounting issues, the public-facing financial reports with the emphasis on traditional linear models of materials flows and asset valuations have thus far not addressed the increased requirement for CE business complexity.

Currently, accounting standards are premised on largely linear business models, which support one-time revenue recognition and matching of costs to revenue, and not having complicated loops of rework on assets that require complex (and subjective) assessments of asset valuation. In order for humankind to flourish in an environment of increasing population and decreasing resources, there will need to be greater emphasis on new ways of working to genuinely recycle materials over much longer time frames, design in longevity and re-use into products and ensure that materials are not contaminated. As a result, new business models, like those predicated on the CE economy will start to emerge.

It may be difficult for investors and capital providers to clearly see the business valuation creation in these new business models, as it may stem from a myriad of sources, rather than just simple sales. This lack of clarity may confuse users of financial reports. In an era where transparency is increasingly valued, this may be a retrograde step, with investors not understanding how a once-simple business might make its money in the future. Accounting standards do need to consider a more 'looped' approach to the return and reuse of assets and raw materials into the business and how both recovered raw materials and remanufactured assets might be reliably measured and monitored. Reporting on these impacts might be better accomplished using less traditional financial reports which stress solely the financial impacts, using instead IR principles instead to explain value creation.

The transition to a CE will require innovative approaches and innovative leaders to manage the complexity of the CE and drive through the radical changes needed to policy and practice (Bebbington & Larrinaga, 2014; Ghisellini *et al.*, 2016). This innovation can come from the accounting profession by changing the perceptions of managers, auditors, investors and policy makers around issues of systems boundaries, systems complexities and interdependencies and the impacts these have on our conventional 'wisdom' of financial reporting of asset values and organisational 'worth'. This could be supported by IR's focus on six capitals, rather than purely financial capital.

Accountants have the 'ear' of those who invest (Gray, 2002) and therefore arguably have a responsibility to change the perceptions of those investors to the broader considerations of society and environment. However, it is impossible for accountants to act alone. To generate the business models of the future will rely on a full 'cradle-to-cradle' systems approach (Braungart & McDonough, 2008). This approach will possess many interconnections and loops in and out of the organisation, which will require a multidisciplinary approach. Whilst some areas of financial reporting have attempted to address the 'business model' approach such as IR (IIRC, 2013), it will require innovative businesses themselves to consider their dependency on the environment as the underpinning for their own longevity and adopt more resource-centric business models. Accountants will need to respond to this by assisting in the process of interpreting these models and their value to others outside the organisation.

REFERENCES

Adams, C., & Frost, G. (2008). Integrating sustainability reporting into management practices. *Accounting Forum*, *32*(4), 288–302. https://doi.org/http://dx.doi.org/10.1016/j.accfor.2008.05.002

Adams, S., & Simnett, R. (2011). Integrated reporting: An opportunity for Australia's not-for-profit sector. *Australian Accounting Review*, *21*(3), 292–301.

Andersen, N. (2016). Blockchain technology a game-changer in accounting? *Deloitte*, 1–5. https://www2.deloitte.com/content/dam/Deloitte/de/Documents/Innovation/Blockchain_A%20game-changer%20in%20accounting.pdf

Anderson, M. S. (2007). An introductory note on the environmental economics of the circular economy. *Sustainability Science*, *2*(1), 133–140. https://doi.org/10.1007/s11625-006-0013-6

Bebbington, J., Brown, J., & Frame, B. (2007). Accounting technologies and sustainability assessment models. *Ecological Economics*, *61*(2–3), 224–236. https://doi.org/10.1016/j.ecolecon.2006.10.021

Bebbington, J., & Larrinaga, C. (2014). Accounting and sustainable development: An exploration. *Accounting, Organizations and Society*, *39*, 395–413. https://doi.org/10.1016/j.aos.2014.01.003

Boulding, K. E. (1966). The economics of coming spaceship earth. In H. Jarret (Ed.), *Environmental quality in a growing economy*. Baltimore: John Hopkins University Press.

Braungart, M., & McDonough, W. (2008). *Cradle to cradle: Remaking the way we make things*. London, UK: Jonathan Cape.

Bringezu, S., Schutz, H., & Mooll, S. (2003). Rational for and interpretation of economy – wide materials flow analysis and derived indicators. *Journal of Industrial Ecology*, *7*(2), 43–64.

Centobelli, P., Cerchione, R., Chiaroni, D., Del Vecchio, P., & Urbinati, A. (2020). Designing business models in circular economy: A systematic literature review and research agenda. *Business Strategy and the Environment*, *29*(4), 1734–1749. https://doi.org/10.1002/bse.2466

Conway, E. (2019). Quantitative impacts of mandatory integrated reporting. *Journal of Financial Reporting and Accounting*, *17*(4), 604–634.

Conway, E., Robertson, F. A., & Ugiagbe-Green, I. (2020). Integrated reporting. In D. Crowther (Ed.), *The palgrave handbook of corporate social responsibility* (p. Forthcoming). London: Palgrave Macmillan.

Eccles, R. G., Krzus, M. P., & Watson, L. A. (2012). Integrated reporting requires integrated assurance. In J. Oringel (Ed.), *Effective auditing for companies: Key developments in practice and procedures* (pp. 161–178). London: Bloomsbury Information.

Ellen Macarthur Foundation (2012a). Circular Economy Case Study – In depth – Washing Machines. Retrieved October 19, 2017, from https://www.ellenmacarthurfoundation.org/circular-economy/interactive-diagram/in-depth-washing-machines

Ellen Macarthur Foundation (2012b). Efficiency vs. Effectiveness in the Circular Economy. Retrieved October 19, 2017, from https://www.ellenmacarthurfoundation.org/circular-economy/interactive-diagram/efficiency-vs-effectiveness

Ellen Macarthur Foundation (2012c). In depth – Mobile Phones – A Circular Economy Study. Retrieved October 19, 2017, from https://www.ellenmacarthurfoundation.org/circular-economy/interactive-diagram/in-depth-mobile-phones

Ellen Macarthur Foundation (2013). The Circular Economy Applied to the Automotive Industry – Ellen MacArthur Foundation. Retrieved October 19, 2017, from https://www.ellenmacarthurfoundation.org/circular-economy/interactive-diagram/the-circular-economy-applied-to-the-automotive-industry

Ellen MacArthur Foundation (2016). Intelligent Assets: Unlocking the Circular Economy Potential. Ellen MacArthur Foundation, 1–25. Retrieved from http://www.ellenmacarthurfoundation.org/assets/downloads/publications/EllenMacArthur Foundation_Intelligent_Assets_080216.pdf

EY (2016). Integrated Reporting: Linking Strategy, Purpose and Value. https://doi.org/10.4324/9781351284646-18

Freeman, R. E., Wicks, A. C., & Parmar, B. L. (2004). Stakeholder theory and "The Corporate Objective Revisited": A reply. *Organization Science, 15*(3), 370–371. https://doi.org/10.1287/orsc.1040.0067

Frishammar, J., & Parida, V. (2019). Circular business model transformation: A roadmap for incumbent firms. *California Management Review, 61*(2), 5–29. https://doi.org/10.1177/0008125618811926

Geissdoerfer, M., Morioka, S. N., de Carvalho, M. M., & Evans, S. (2018). Business models and supply chains for the circular economy. *Journal of Cleaner Production, 190*, 712–721. https://doi.org/10.1016/j.jclepro.2018.04.159

Geng, Y., & Doberstein, B. (2008). Developing the circular economy in China: Challenges and opportunities for achieving "leapfrog" development. *International Journal of Sustainable Development and World Ecology, 9*(4), 333–340.

Geng, Y., Fu, J., Sarkis, J., & Xue, B. (2012). Towards a national circular economy indicator system in China: An evaluation and critical analysis. *Journal of Cleaner Production, 23*(1), 216–224. https://doi.org/10.1016/j.jclepro.2011.07.005

Geng, Y., Zhu, Q., Doberstein, B., & Fujita, T. (2008). Implementing China's circular economy concept at the regional level: A review of progress in Dalian, China. *Waste Management 29*. https://doi.org/10.1016/j.wasman.2008.06.036

Ghisellini, P., Cialani, C., & Ulgiati, S. (2016). A review on circular economy: The expected transition to a balanced interplay of environmental and economic systems. *Journal of Cleaner Production, 114*, 11–32. https://doi.org/10.1016/j.jclepro.2015.09.007

Gray, R. (2002). Of messiness, systems and sustainability: Towards a more social and environmental finance and accounting. *British Accounting Review, 34*(4), 357–386.

Gray, R. (2006). Does sustainability reporting improve corporate behaviour? Wrong question? *Right Time? Accounting and Business Research, 36*, 65–88.

Gray, R., & Bebbington, J. (2001). *Accounting for the environment.* London: Sage.

Hedgepeth, W. O. (2006). *RFID metrics: Decision making tools for today's supply chains.* Boca Raton, USA: CRC/Taylor & Francis.

Hopkinson, P., Zils, M., Hawkins, P., & Roper, S. (2018). Managing a complex global circular economy business model: Opportunities and challenges. *California Management Review, 60*(3), 71–94. https://doi.org/10.1177/0008125618764692

Hu, J., Xaio, Z., Deng, W., & Ma, S. (2011). Ecological utilization of leather tannery waste with circular economy model. *Journal of Cleaner Production, 19*, 221–228.

IASB (2016a). IAS 2 Inventories.

IASB (2016b). IFRS 16 Leases At a glance.

IIRC (2011). Communicating Value in the 21st Century. Retrieved from www.theiirc.org

IIRC (2013). The International Integrated Reporting framework. *The International <IR> Framework.*

IIRC (2019a). Get to grips with the six capitals. Retrieved October 21, 2019, from https://integratedreporting.org/what-the-tool-for-better-reporting/get-to-grips-with-the-six-capitals/

IIRC (2019b). What? The tool for better reporting. Retrieved October 21, 2019, from https://integratedreporting.org/what-the-tool-for-better-reporting/

Lahane, S., Kant, R., & Shankar, R. (2020). Circular supply chain management: A state-of-art review and future opportunities. *Journal of Cleaner Production, 258*, 120859. https://doi.org/10.1016/j.jclepro.2020.120859

Larrinaga-González, C., Carrasco-Fenech, F., Javier, F., Carmen, C.-G., José, C.-R., Páez-Sandubete, M., & Páez, M. (2001). The role of environmental accounting in organizational change – An exploration of Spanish companies. *Accounting, Auditing & Accountability Journal, 14*(2), 213–239. Retrieved from http://dx.doi. org/10.1108/09513570110389323

Larrinaga, C., Carrasco, F., Correa, C., Llena, F., & Moneva, J. (2002). Accountability and accounting regulation: The case of the Spanish environmental disclosure standard. *European Accounting Review, 11*(4), 723–740. https://doi.org/10.1080/0963818022000001000

Lieder, M., & Rashid, A. (2016). Towards circular economy implementation: A comprehensive review in context of manufacturing industry. *Journal of Cleaner Production, 115*, 36–51. https://doi.org/10.1016/j.jclepro.2015.12.042

Lüdeke-Freund, F., Gold, S., & Bocken, N. M. P. (2019). A review and typology of circular economy business model patterns. *Journal of Industrial Ecology, 23*(1), 36–61. https://doi.org/10.1111/jiec.12763

Murray, A., Skene, K., & Haynes, K. (2017). The circular economy: An interdisciplinary exploration of the concept and application in a global context. *Journal of Business Ethics, 140*(3), 369–380. https://doi.org/10.1007/s10551-015-2693-2

Nylander, J. (2015). Why "Integrated Reporting" Attracts Investors. Retrieved December 3, 2018, from https://www.forbes.com/sites/jnylander/2015/11/11/why-integrated-reporting-attracts-investors/#10dcc87257e7

Pedersen, E. R. G., Earley, R., & Andersen, K. R. (2019). From singular to plural: Exploring organisational complexities and circular business model design. *Journal of Fashion Marketing and Management, 23*(3), 308–326. https://doi.org/10.1108/JFMM-04-2018-0062

Percy, S. (2016). Building Blocks. *Americas*, 5. Retrieved from http://www.ey.com/Publication/vwLUAssets/EY-building-blocks-of-the-future/$FILE/EY-building-blocks-of-the-future.pdf

Pieroni, M. P. P., McAloone, T. C., & Pigosso, D. C. A. (2019). Configuring new business models for circular economy through product-service systems. *Sustainability, 11*(13). https://doi.org/10.3390/su11133727

Rinaldi, L., Unerman, J., & de Villiers, C. (2018). Evaluating the integrated reporting journey: insights, gaps and agendas for future research. *Accounting, Auditing and Accountability Journal, 31*(5), 1294–1318. https://doi.org/10.1108/AAAJ-04-2018-3446

Rosa, P., Sassanelli, C., & Terzi, S. (2019). Circular business models versus circular benefits: An assessment in the waste from electrical and electronic equipments sector. *Journal of Cleaner Production, 231*, 940–952. https://doi.org/10.1016/j.jclepro.2019.05.310

Rudnicka, A. (2018). Business models in circular economy concept. *Prace Naukowe Uniwersytetu Ekonomicznego We Wrocławiu, 520*, 106–114. https://doi.org/10.15611/pn.2018.520.09

Singh, J., & Ordoñez, I. (2016). Resource recovery from post-consumer waste: Important lessons for the upcoming circular economy. *Journal of Cleaner Production, 134*, 342–353. https://doi.org/10.1016/j.jclepro.2015.12.020

Stahel, W. R. (1991). *Langlebigkeit und material recycling : Strategien zur vermeidung von abfällen im bereich der produkte.* Essen: Vulkan-Verlag. Retrieved from http://www.product-life.org/en/archive/case-studies/washing-machines

Teece, D. J. (2010). Business models, business strategy and innovation. *Long Range Planning, 43*(2–3), 172–194.

Tukker, A. (2015). Product services for a resource-efficient and circular economy – A review. *Journal of Cleaner Production, 97*, 76–91. https://doi.org/10.1016/j.jclepro.2013.11.049

Tukker, A., Eder, P., & Suh, S. (2006). Environmental impacts of products. *Journal of Industrial Ecology, 10*(3), 183–198. Retrieved from http://search.ebscohost.com/login. aspx?direct=true&db=a9h&AN=21615230&site=ehost-live

UK Government. (2018). The Waste Electrical and Electronic Equipment (Amendment) Regulations 2018. Retrieved July 9, 2020, from http://www.legislation.gov.uk/uksi/2018/102/made

Ünal, E., Urbinati, A., & Chiaroni, D. (2019). Managerial practices for designing circular economy business models: The case of an Italian SME in the office supply industry. *Journal of Manufacturing Technology Management, 30*(3), 561–589. https://doi.org/10.1108/JMTM-02-2018-0061

Urbinati, A., Chiaroni, D., & Chiesa, V. (2017). Towards a new taxonomy of circular economy business models. *Journal of Cleaner Production, 168*, 487–498. https://doi.org/10.1016/j.jclepro.2017.09.047

Section 2

Achieving Sustainability through Circular Economy Practices

4 Sustainability Through Green Manufacturing Systems
An Applied Approach

Mahender Singh Kaswan and Rajeev Rathi
School of Mechanical Engineering,
Lovely Professional University, Phagwara, Punjab, India

Ammar Vakharia
School of Electrical and Electronics Engineering, Lovely
Professional University, Phagwara, Punjab, India

CONTENTS

4.1 INTRODUCTION

Over the last few decades, due to intense competition, shortage of resources and sustainable oriented customers' demand, the industrial organizations are under tremendous pressure to improve their environmental and societal performances (Kumar *et al.*, 2020; Kaswan & Rathi, 2020b). The industries can get the competitive edge by offering the green products to the customers, as in the modern era of sustainability the customers have exhibited proclivity towards ecofriendly products (Khan *et al.*, 2020). Nearly, 80% of the global demand for the energy is fulfilled by fossil fuel–based energy methods and it is expected to increase in future (Mangla *et al.*, 2020). The industries are facing a bunch of challenges to sustain in the globalized market. To remain competitive in the global market, industries have to shift their current methods of doing work towards sustainable ones that not only ensures the environmental sustainability, but also leads to enhanced productivity and profitability (Kumar *et al.*, 2019; Singh *et al.*, 2020). Table 4.1 indicates the societal, economic and environmental challenges of the modern enterprise. To mitigate these challenges, there is an immense need for sustainable development approaches that meet the demand of the present without compromising the ability of future generations to meet their own needs (Pandey *et al.*, 2018; Kumar *et al.*, 2020).

The products after their primary use are either buried or burned that leads to the emission of the harmful gases in the environment (Hussain *et al.*, 2019). The sustainable approaches use the concepts of reduce, reuse, recycle (3 R's) and renewable methods to combat with environmental-related emissions (Khan *et al.*, 2020; Kaswan & Rathi, 2020a). Figure 4.1 depicts the traditional life cycle of a product that is directly disposed of in the land after its prime use by the end customer. So, there is the necessity to improve the traditional life cycle to the new product life cycle in which the product after primary use is either used as a ready-made raw material for

TABLE 4.1
Challenges to Manufacturing

S. No.	Economical	Environmental	Societal
1	Shorter product life cycle	Climate change	Aging workforce
2	Technological changes	Depletion of the natural resources	Non-availability of the skilled workforce
3	Demand fluctuation		
4	Product variety		

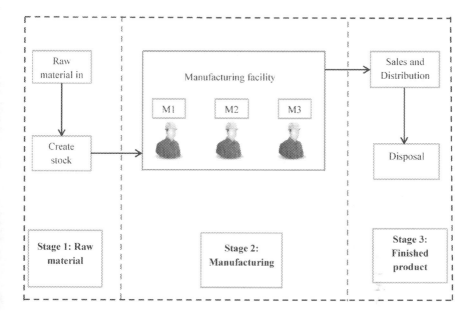

FIGURE 4.1 Stages in the product life cycle (production and sales view).

another industry or recycled. The industrial organization should focus on the identification and reduce the environmental impact at each stage of the lifecycle of the
product (Kaswan & Rathi, 2019; Siegel *et al.*, 2019). The main motive to identify and
measure the environmental impact is that once we can measure the emission, one
can find out the ways to mitigate the same (Sony and Naik, 2019).

This chapter has been divided into five sections, including the introduction.
Section 2 of the manuscript is dedicated to the life cycle assessment (LCA).
Section 3 of this chapter focuses on the basics of green manufacturing (GM): indicators, development of GM system, etc. The practices of GM have demonstrated
in Section 4 of this chapter. The final segment of this work reflects the conclusion
drawn from the comprehensive discussion on GM.

4.2 LIFE CYCLE ASSESSMENT

LCA is a tool used for a thorough examination of a product or a process life cycle,
from the initial stage to the disposal, in order to identify, measure and improve the
impact it has on the environment (Kaswan *et al.*, 2020). LCA had the first application
in packaging studies and in the early stages of its development, mainly focused on
energy and emissions (Agrawal and Vinodh, 2020). To date, many companies have
taken the initiative to reduce their carbon footprint by using the LCA method. Toppan
printing company, based in Japan, has used the LCA method in order to understand
their impact on the environment. They have been very successful in reducing their
carbon footprint as can be evident from their annual open-access report where they
mentioned one of the current aims for the year 2020 is to reduce their carbon emissions from 751 (in 2008) to 551 kilotons at the domestic sites (Toppan Environmental

Performance Data, 2019). While a study conducted in Denmark on LCA of wind farms found the energy payback time of land-based wind farms to be nearly three months and for offshore wind farms, nearly four and a half months (Kaswan *et al.*, 2020). Similarly, a recent study suggested that the carbon footprint of the Danish wind turbine fleet will be reduced from 40 to 13 g CO_2-eq/kWh by the year 2030 (Besseau, 2019).

The first step of LCA deals with the comparison of the outlining of the goal and scope of the study. The goal is related to expressing the improvements that one wants to make in the existing process or established system (Lv *et al.*, 2019). The scope is defining the boundary of the study being considered. Here, it is decided that whether you are considering a unit process, the entire system, or the project for study. Once the goal and scope of the study have been defined, the next step is the analysis of various inventories. Figure 4.2 depicts lifecycle assessment procedure. The inventory analysis is done on the data set collected related to process energy, feedstock energy, wastes and pollutants from the input stage to the finished goods.

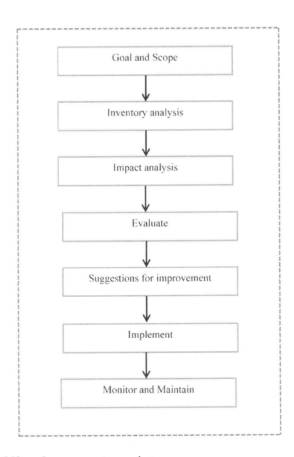

FIGURE 4.2 Life cycle assessment procedure.

The next step of LCA is impact analysis; here, GHGs emissions, etc. are found from the existing process or system under consideration. The evaluation of the various wastes and emission potentials is made in the following LCA step. The possible green and sustainable solutions are proposed in the next stage of the LCA. The best solution is implemented in the subsequent stage of the LCA for reduced environmental-related emissions. The findings after the implementation of the potential solution are monitored in the next step, and if substantial improvements are reported, then the adopted solution or methods is sustained.

4.3 GREEN MANUFACTURING

The GM is a method that transforms the traditional environmental damaging manufacturing process to a more efficient and eco-friendly process (Dornfeld, 2012; Leong, 2019). The GM system is such a kind manufacturing system that provides high material, energy and material efficiencies through the advanced methods of production like dry machining and cryogenic machining (Rusinko, 2007).

4.3.1 GREEN MANUFACTURING INDICATORS

GM indicators represent the state or level of something (Deif, 2011). Metrics are the yardsticks or parameters or variables. Indicators may have further metrics with it (Sun et al., 2020). Table 4.2 indicates various indicators of GM at multiple stages of product development.

TABLE 4.2
Indicators of Green Manufacturing

S. No.	Input	Operations	Products
1	Nonrenewable material intensity	Water intensity	Recycle/ reuse content
2	Restrictive substance content	Energy intensity	Recyclability
3	Recycle and renewable material content	Renewable proportion intensity	Renewable material content
4		GHGs intensity	Nonrenewable material intensity
5		Residual intensity	Restricted substance content
6		Air release intensity	Energy consumption intensity
7		Water release intensity	GHGs intensity
8		Proportional of natural land	

4.3.2 Metrics of Green Manufacturing

Sustainability covers three dimensions, i.e. social, economic and environmental to make a resilient society. There are three types of sustainability metrics (Dornfeld, 2012).

1. Economic
2. Environmental
3. Societal

4.3.2.1 Economic Metrics

a. Payback period
   The payback period is the time after which all the initial investment is recovered through the revenue.
b. Internal rate of revenue (IRR)
   The internal rate of revenue gives the annual effective compound rate of investment.
c. Net present value
   The net present value determines whether an investment in future cost saving is worthwhile, and if it requires adjusting future cash flow into a current time frame.
d. Cost of ownership (COO)
   The cost of the ownership is calculated using Equation (1)

$$COO = \frac{C_1 + C_2 + \frac{C_3}{(1+i)Z}}{B \times V_E} \times N \tag{1}$$

Where, C_1 = Equipment cost
C_2 = Set up cost
C_3 = Annual operational cost
I = discounting rate
Z = years into future
B = amount of output
V_E = Economic value of unit output

The setup cost is estimated using Equation (2)

$$C_2 = I + T + \sum P \times F \tag{2}$$

Here, I = installation fee
T = transportation cost

The annual cost of operation is calculated using the Equation (3)

$$C_3 = S \times R_S + U \times R_E + \sum O + H \times R_m \tag{3}$$

S = footprint
R_S = footprint cost or rate
U = energy/electricity
R_E = rate of electricity
O = consumable cost
H = downtime in hours
R_m = maintenance cost/hour

4.3.2.2 Economic Metrics

The following are the environmental metrics:

a. Utility analysis
b. Single quantity analysis
c. Hierarchical analysis
d. Verbal argument approach

The environmental metrics are also classified based on the aggregation method

i. Low level: CO_2, GHGs, global warming
ii. Air, land, water

4.3.2.3 Societal Metrics

a. Internal dissatisfaction (inside the workers)
b. External

It considers the overall community. It covers employees, shareholders, consumers and the local community.

4.3.3 PRODUCT LEVEL INDICATOR OF GM

4.3.3.1 Carbon Footprint

It is the amount of emission of CO_2/CH_4 for a defined population/system/activity/product (Kong *et al.*, 2020). It includes all the sources, sink and storage. Here the spatial constraints and temporal boundaries are decided.

4.3.3.2 Ford Product Sustainability Index

It is the index used for the automobile industry and even for service organizations. It looks at eight different variables across three pillars (Table 4.3)

4.3.4 INDICES FOR GREEN MANUFACTURING (INDUSTRY LEVEL)

4.3.4.1 Eco-compass (Fuller and James, 1996)

It is dependent on LCA data. It included six dimensions.

a. Energy intensity
b. Mass intensity

TABLE 4.3
Ford Product Sustainability Index

S. No.	Environmental	Economic	Social
1	Life cycle global warming potential	Life cycle cost	Mobility capability
2	Life cycle air quality potential	Cost of ownership	Safety
3	Sustainable material		
4	Restricted substance		
5	Drive (noise)		

c. Health and potential environmental risk
d. Resource conservation
e. Extent of reuse/remanufacturing/recycling
f. Service extension

4.3.4.2 Eco-indicator 99

It is a state of the art of the damage oriented life cycle impact assessment, i.e. it identifies the damage to the environment.

4.3.4.3 Green Co Rating System (CII, 2011)

It gives point to the industry based on the various indicators. Here, the total points to be considered are 1000. Figure 4.3 indicates the various levels of the green co rating system.

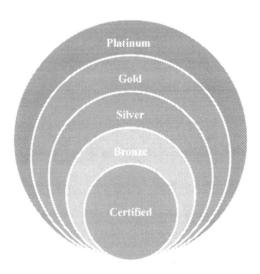

FIGURE 4.3   Green co rating levels.

TABLE 4.4
Number of Indicators to be Used

S. No.	Experience Level	No. of Indicators to Select	Basis of Indicator Selection
1	Beginner	1–5	Data already available collected
2	Intermediate	6–12	Priorities highlighted through your issue identification
3	Advanced level	13–18+	All indicators are relevant to the facility. Additional indicators may be developed to facilitate further improvement

4.3.4.4 Number of Indicators to Use (OECD, 2011)

The organization for economic co-operation and development provides the guidelines for the number of indicators to be used for the development of the GM system within the organization. Table 4.4 depicts the number of indicators to be used for adopting a GM system.

4.3.5 Green Manufacturing Development Procedure

There are four significant steps in the development of the GM system.

a. **Identification of the current state of the system under consideration**
   The first step towards the development of the GM system is the assessment of the current state of the system under consideration (Figure 4.4). The level of Lean and Green wastes, environmental and other associated emissions are measured. The tools used at this stage may be a structurally designed

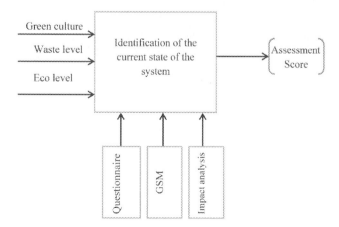

FIGURE 4.4 Assessment of current system state.

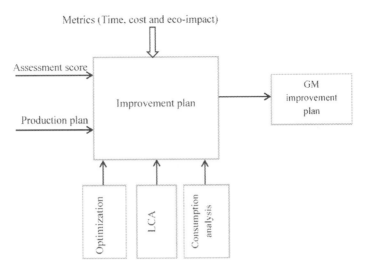

FIGURE 4.5   Improvement plan.

set of questionnaires, green stream mapping (GSM), or impact analysis. Finally, an assessment score is calculated that determines the current level of the greenness of the system under consideration.

b. **Develop the improvement plan**

The next step in the development of the GM system is the development of an improvement plan (Figure 4.5). Here, based on the assessment score and production plan, improvement measures are made. The optimizations of the various variables are made, LCA is made, and consumption analysis is carried out to find out the best improvement plan. The plan here must make a strategic impact on all the dimensions of the sustainability of the concerned organization.

c. **Apply the implementation plan**

The third step in the development of the GM system is introduction of an implementation plan (Figure 4.6). The improvement plan for increased material, energy efficiency and productivity are applied at this stage. The various metrics are made for finding the sustainability content of the system under consideration. The parameters considered here are recyclable material content, restricted substance content, GHGs intensity, renewable material content, energy intensity, water intensity, etc. The implemented plan resulted in the increased utilization and efficiency of the machine of the system, improved process, and increased overall efficacy and sustainability of the system under consideration.

d. **Maintain**

The next step in the development of a GM system is to maintain or sustain or control the adopted improvement plan for increased sustainability of the system being considered (Figure 4.7). The improvement in the system in terms of increased financial capabilities, productivity, and

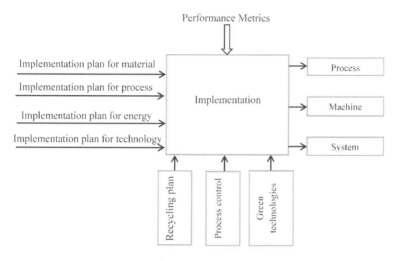

FIGURE 4.6   Implementation plan.

improved levels of environmental emissions are measured from time to time so that capacity of the adopted plan can be judged and if there is any distraction plan be modified accordingly. Meanwhile, small pursuits are always made with the selected method for improved societal, economic and environmental dimensions of sustainability

4.4   GREEN MANUFACTURING PRACTICES

GM is such a form of production that leads to less carbon footprint and produces environmentally friendly products (Bhanot, 2017). The subsequent subsections depict some of the practices in the era of GM that only serves the mother nature but also lead to increased profitability dynamics in the long run.

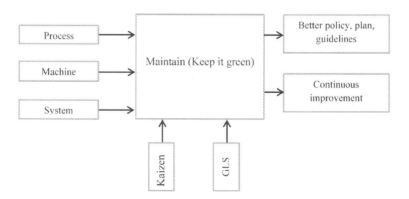

FIGURE 4.7   Maintenance of the adopted green manufacturing plan.

4.4.1 INDUSTRIAL SYMBIOSIS

Symbiosis means a local partnership, where it provides, share and reuse of resources to generate shared value (Chertow, 2007). The industrial symbiosis is a concept where two companies make an association in which waste or by-product of one is used as input by the other (Jacobsen, 2006). Figure 4.8 reprsents different motives to have industrial symbiosis. Industrial symbiosis is cooperation between two or more separate industries to exchange resources, by products or materials that are no longer of use to one, but of use to other (Baldassarre, 2019). It is beneficial from commercial as well as an environmental viewpoint. The application of this concept allows materials to be used more sustainably and contributes to the creation of a circular economy. It leads to reduction in the requirement of raw material, reduction of waste of associated carbon footprints (Cao et al., 2020). Moreover, industrial symbiosis generates new business opportunities for recyclers and trading companies.

The depletion of the resources and increased cost of the products have forced the organization to look for the circular concepts of the production (Patala et al., 2020). The strategic partnership among the partners through sharing of material, energy, water streams, etc. builds the resilience, increases monetary benefits and leads to a healthy environment. The success of the symbiosis lies from the Kalundborg symbiosis (1960s), where all the partners collaborated and lead to a substantial reduction in material usage and GHGs emissions.

4.4.2 DRY MACHINING

The increased share of the cutting fluid in the cost of the product and environmental concerns associated with its use led to the development of the dry

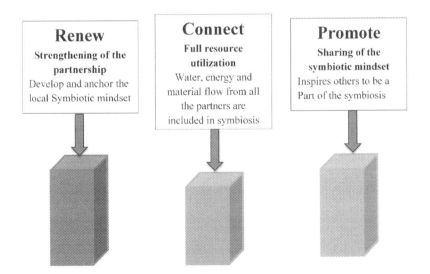

FIGURE 4.8 Motives of the indutrial symbiosis.

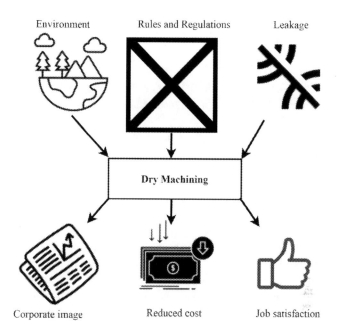

Environment   Rules and Regulations   Leakage

Corporate image   Reduced cost   Job satisfaction

FIGURE 4.9 Dry machining.

machining process. In the dry machining process, no cutting fluid is used. The cutting fluid contributes nearly 15–20% of the total cost of the workpiece (Sreejith and Ngoi, 2000). It should be noted that a few of the benefits of the cutting fluid would not be available in the case of the dry machining. Figure 4.9 depicts the need and potential benefits that can be drawn from the dry machining. The increased levels of greenhouse gases have resulted in deteriorating human health and work ambiance. Moreover, the intergovernmental policies, international treaties like the Paris agreement have enforced the industrial organization to cut the current level of emissions. The leakage of the cutting fluid in the conventional type of machining and their disposal further leads to the negative environmental impacts (Chaudhary & Mudimela, 2020; Devillez *et al.*, 2011; Kumar and Kaswan, 2016). So, the need for such machining that uses no coolants, and at the same environmentally friendly, was felt. The dry machining posits the answers to these questions. The organizations that deliver eco-friendly or sustainable products have a better corporate image. The organizations that adopt green practices not only get increased productivity but have enhanced profitability in the long run (Chaudhary & Mudimela, 2020; Derflinger *et al.*, 1999; Nouari *et al.*, 2003). The organizations that adapt the green practices will not only benefit the environment by reducing raw material consumption, but their product quality and productivity are also enhanced (Jia & Wang, 2019). Moreover, the organization's members feel boosted with the production of sustainable products that not only serve the purpose of the society but also bring increased productivity and profitability in the organization (Kaswan *et al.*, 2019).

The indirect contact with the lubricant of cutting fluid has headed to the following advancements in the era of dry machining:

- The cutting fluid flows through a well-designed channel below the insert of the main cutting zone (Ber & Goldblatt, 1989; Fukui, 2004).
- The vaporized fluid enters through the shanks and cools the cutting zone (Jeffries, 1969).
- Cryogenic cooling: very low temperature enters through the well-deliberated channel and cools the cutting area.
- Use of the thermoelectric elements through the formation of the cooling zone (Meyers, 1964).
- Make the tool material more refractory, and advancement in the cutting tool material has led to phenomenal success in the era of dry machining.

The dry machining is not suitable for all sets of materials like titanium and nickel. But with the MoS_2 solid lubricant can be injected near the rake and flank face of the tool through the small hole. The dry machining also leads to the high temperature of the workpiece that subsequently change the structural properties of the material. The continuous development in the field of the dry machining led to near dry machining (NDM) or minimum quantity lubrication machining.

4.4.3 NEAR DRY MACHINING

The cutting fluids are used in the machining process to carry chips, increase tool life and reduce the temperature of the cutting zone. But the disposal of the cutting fluid is the primary concern as it leads to skin diseases, eye rashes, cardiovascular diseases, etc. The cutting fluid is termed as the necessary evil because it leads to harm to human health, but it is essential for the machining process. The dry machining leads to cleaner parts, no waste, reduced machining cost and reduced recycling cost. But dry machining demands a considerable investment. The machines and tools that are adaptable with cutting fluids cannot be used for dry machining. DM requires high capacity and power machines and specially designed tools and fixtures that can withstand high temperatures. The distortion results in the machined parts due to the high cutting force and thermal expansion. Furthermore, handling and gauging of the parts produced by dry machining pose severe problems as the manufactured parts are quite hot. So, these disadvantages associated with DM lead to subsequent research and development in the field of machining, and consequently, NDM comes into the picture. NDM reduces tool wear, cutting zone temperatures, friction coefficients, as well as improves surface finish as compared to dry machining (Sharma et al., 2016). Figure 4.10 depicts near dry machining.

So, to reduce the effects on human and ecological health in the constant search for dry machining, a process that provides the practical solution called NDM was developed. The process uses a minimum quantity of lubricant or cutting fluid and provides maximum performance (Sharma & Sidhu, 2014). NDM uses a mixture of air and oil termed as an aerosol for the lubrication purpose in the machining zone (Astakhov, 2008). NDM leads to improved surface finish and enhanced cutting tool

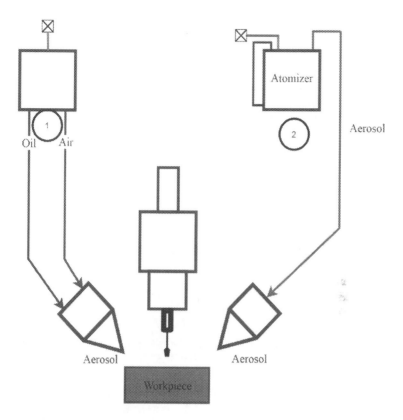

FIGURE 4.10   Near dry machining.

life as compared to its counterpart, dry machining (Li & Liang, 2007). NDM uses a minimal quantity of the cutting fluid in the machining zone. The aerosol is considered as the suspension of solid and liquid into the air. The oil is dispersed into the air in the NDM for the lubrication purpose (Fratila, 2009). The liquid is converted into the small droplets or mist by a process called atomization in which the liquid is passed through the specially designed set of nozzles. The NDM can be classified according to the type of the aerosol supply:

NDM 1: NDM with external aerosol supply. In NDM 1, the aerosol is supplied by an outer nozzle placed in the machine (Kao *et al.*, 2007).
NDM 2: NDM with internal (through-tool) aerosol supply. In NDM 2, the aerosol is supplied through the tool.

NDM 1.1 with an ejector nozzle
The oil and the compressed air are supplied to the ejector nozzle, and the aerosol is formed just after the injector.
NDM 1.2 with a conventional nozzle
The aerosol is prepared in an external atomizer and then supplied to a standard nozzle. The nozzle design is similar to that used in flood MWF supply.

Advantages of NDM 1

- The cost of the equipment is low as compared to another version of NDM.
- The same type of tool can be used for flood metalworking fluid.
- The ease of operation and maintenance.
- The used equipment is portable.
- To suit the requirements of the various types of machining and toolset, different types of nozzles and toolset are available.

Disadvantages of NDM 1

- NDM 1.1 and NDM 1.2 are not suitable or drilling and boring purposes.
- The position of the nozzle is fixed concerning the movement of the parts.
- The NDM 1.1 and 1.2 are not suitable for the job shop machining as the aerosol parameters have to be adjusted for each type of machining.

4.4.4 CRYOGENIC MACHINING

The machining process that is performed below 123°K is called as cryogenic machining (Wang & Rajurkar, 2000; Grguraš et al., 2019). In this type of machining, a cryogenic fluid (liquid nitrogen) is used. The boiling point of the liquid nitrogen is -195°C or 77°K. The liquid nitrogen is carried in a well-insulated vacuum flask (Hong & Ding, 2001). In the past, the drawback associated with cryogenic machinic was that liquid nitrogen gets vaporized before reaching into the tool-workpiece contact area (Narayanan & Jagadeesha, 2020). To overcome this, vacuum jacketed feed lines deliver the liquid nitrogen at a slow rate, and liquid nitrogen doesn't get vaporized. The word cryo stands for frost, and genic means to produce. In this machining process, the liquid nitrogen is delivered to the cutting area through the spindle-tool very near to the cutting edge.

4.4.4.1 Motivation Behind Cryogenic Machining
- Reduces the toxicity in the factory.
- Reduces the harmful bacterias.
- Reduces the cost of the disposal.
- It enhances the tool life nearly ten times as compared to conventional counterparts.
- The processing speed of the machining is increased as increased cooling allows the tool to operate at high speed.

4.5 CONCLUSION

The increased competition and environmental policies on climate change have brought immediate attention to the industrial managers to shift methods of doing business towards sustainable ones. GM is that form of production that produces the eco-friendly products in harmony with mother nature. The industry must change traditional practices and made a GM system that not only generates quality products but with less carbon footprint. The manufacturing organization is widely shifting its conventional methods to GM methods like NDM, cryogenic machining, etc. The industrial organization must focus on the concept of the industrial symbiosis where the wastes or other disposal materials are used as raw material by another industry. So, the ideas of the circular economy, sharing economy, industry 4.0 are supplemented with the popular notion of GM.

REFERENCES

Agrawal, R., & Vinodh, S. (2020). Life Cycle Assessment of an Additive Manufactured Automotive Component. In *Advances in additive manufacturing and joining* (pp. 219–228). Singapore: Springer.

Astakhov, V. P. (2008). Ecological Machining: Near-Dry Machining. In *Machining* (pp. 195–223). London: Springer.

Baldassarre, B., Schepers, M., Bocken, N., Cuppen, E., Korevaar, G., & Calabretta, G. (2019). Industrial symbiosis: Towards a design process for eco-industrial clusters by integrating circular economy and industrial ecology perspectives. *Journal of Cleaner Production, 216*, 446–460.

Ber, A., & Goldblatt, M. (1989). The influence of temperature gradient on cutting tool's life. *CIRP Annals, 38*(1), 69–73.

Besseau, R., Sacchi, R., Blanc, I., & Perez-Lopez, P. (2019). Past, present and future environmental footprint of the Danish wind turbine fleet with LCA_WIND_DK, an online interactive platform. *Renewable and Sustainable Energy Reviews, 108*, 274–288.

Bhanot, N. (2017). *Development of an integrated sustainable manufacturing assessment framework for turning process,* Doctoral dissertation.

Cao, X., Wen, Z., Xu, J., De Clercq, D., Wang, Y., & Tao, Y. (2020). Many-objective optimization of technology implementation in the industrial symbiosis system based on a modified NSGA-III. *Journal of Cleaner Production, 245*, 118810.

Chaudhary, R., & Mudimela, P. R. (2020). Pull-in response and eigen frequency analysis of graphene oxide-based NEMS switch. *Materials Today: Proceedings.* doi.org/10.1016/j.matpr.2020.01.560.

Chaudhary, R., & Mudimela, P. R. (2020). 3D modeling of graphene oxide based nano electromechanical capacitive switch. *Microsystem technologies micro- and nanosystems-information storage and processing systems.*

Chertow, M. R. (2007). "Uncovering" industrial symbiosis. *Journal of Industrial Ecology, 11*(1), 11–30.

Deif, A. M. (2011). A system model for green manufacturing. *Journal of Cleaner Production, 19*(14), 1553–1559.

Derflinger, V., Brändle, H., & Zimmermann, H. (1999). New hard/lubricant coating for dry machining. *Surface and Coatings Technology, 113*(3), 286–292.

Devillez, A., Le Coz, G., Dominiak, S., & Dudzinski, D. (2011). Dry machining of Inconel 718, workpiece surface integrity. *Journal of Materials Processing Technology, 211*(10), 1590–1598.

Dornfeld, D. A. (Ed.) (2012). *Green manufacturing: Fundamentals and applications.* New York: Springer Science & Business Media.

Ferreira, I. A., Barreiros, M. S., & Carvalho, H. (2019). The industrial symbiosis network of the biomass fluidized bed boiler sand mapping its value network. *Resources, Conservation and Recycling, 149*, 595–604.

Fratila, D. (2009). Evaluation of near-dry machining effects on gear milling process efficiency. *Journal of Cleaner Production, 17*(9), 839–845.

Fukui, H., Okida, J., Omori, N., Moriguchi, H., & Tsuda, K. (2004). Cutting performance of DLC coated tools in dry machining aluminum alloys. *Surface and Coatings Technology, 187*(1), 70–76.

Grguraš, D., Sterle, L., Krajnik, P., & Pušavec, F. (2019). A novel cryogenic machining concept based on a lubricated liquid carbon dioxide. *International Journal of Machine Tools and Manufacture, 145*, 103456.

Hong S. Y., & Ding, Y. (2001). Cooling approaches and cutting temperatures in cryogenic machining of Ti-6Al-4V. *International Journal of Machine Tools and Manufacture, 41*(10), 1417–1437.

Huntzinger, D. N., & Eatmon, T. D. (2009). A life-cycle assessment of Portland cement manufacturing: Comparing the traditional process with alternative technologies. *Journal of Cleaner Production, 17*(7), 668–675.

Hussain, K., He, Z., Ahmad, N., & Iqbal, M. (2019). Green, lean, six sigma barriers at a glance: A case from the construction sector of Pakistan. *Building and Environment, 161*, 106225.

Jacobsen, N. B. (2006). Industrial symbiosis in Kalundborg, Denmark: A quantitative assessment of economic and environmental aspects. *Journal of Industrial Ecology, 10*(1–2), 239–255.

Jeffries, N. P. (1969). *A new cooling method for metal-cutting tools,* Doctoral dissertation, University of Cincinnati.

Jia, X., & Wang, M. (2019). The Impact of Green Supply Chain Management Practices on Competitive Advantages and Firm Performance. In *Environmental sustainability in Asian logistics and supply chains* (pp. 121–134). Singapore: Springer.

Kao, C. C., Tao, J., & Shih, A. J. (2007). Near dry electrical discharge machining. *International Journal of Machine Tools and Manufacture, 47*(15), 2273–2281.

Kaswan, M. S., & Rathi, R. (2019). Analysis and modeling the enablers of green lean six sigma implementation using interpretive structural modeling. *Journal of Cleaner Production, 231*, 1182–1191.

Kaswan, M. S., & Rathi, R. (2020a). Investigating the enablers associated with implementation of Green Lean Six Sigma in manufacturing sector using best worst method. *Clean Technologies and Environmental Policy, 22*, 865–876. https://doi.org/10.1007/s10098-020-01827-w.

Kaswan, M. S., & Rathi, R. (2020b). Green lean six sigma for sustainable development: Integration and framework. *Environmental Impact Assessment Review, 83*, 106396.

Kaswan, M. S., Rathi, R., Khanduja, D., & Singh, M. (2020). Life Cycle Assessment Framework for Sustainable Development in Manufacturing Environment. In *Advances in intelligent manufacturing* (pp. 103–113). Singapore: Springer.

Kaswan, M., Rathi, R., & Singh, M. (2019). Just in time elements extraction and prioritization for health care unit using decision making approach. *International Journal of Quality & Reliability Management, 36*(7), 1243–1263.

Khan, S. A. R., Zhang, Y., Kumar, A., Zavadskas, E., & Streimikiene, D. (2020). Measuring the impact of renewable energy, public health expenditure, logistics, and environmental performance on sustainable economic growth. *Sustainable Development.* Vol 28 No. 4, 833–843 doi.org/10.1002/sd.2034.

Kong, M., Pei, J., Liu, X., Lai, P. C., & Pardalos, P. M. (2020). Green manufacturing: Order acceptance and scheduling subject to the budgets of energy consumption and machine launch. *Journal of Cleaner Production, 248*, 119300.

Kumar, A., Choudhary, S., Garza-Reyes, J. A., Kumar, V., Khan, S. R. A., & Mishra, N. (2020). Analysis of critical success factors for implementing industry 4.0 integrated circular supply chain–Moving towards sustainable operations. Production Planning and Control.

Kumar, A., Moktadir, A., Liman, Z. R., Gunasekaran, A., Hegemann, K., & Khan, S. A. R. (2020). Evaluating sustainable drivers for social responsibility in the context of ready-made garments supply chain. *Journal of Cleaner Production, 248*, 119231.

Kumar, A., Zavadskas, E. K., Mangla, S. K., Agrawal, V., Sharma, K., & Gupta, D. (2019). When risks need attention: Adoption of green supply chain initiatives in the pharmaceutical industry. *International Journal of Production Research, 57*(11), 3554–3576.

Kumar, M., & Kaswan, M. S. (2016). Optimization of surface roughness & MRR in end milling on D2 steel using Taguchi method. *OPTIMIZATION, 5*(1), 4–9.

Leong, W. D., Lam, H. L., Ng, W. P. Q., Lim, C. H., Tan, C. P., & Ponnambalam, S. G. (2019). Lean and green manufacturing—a review on its applications and impacts. *Process Integration and Optimization for Sustainability, 3*(1), 5–23.

Li, K. M., & Liang, S. Y. (2007). Modeling of cutting forces in near dry machining under tool wear effect. *International Journal of Machine Tools and Manufacture, 47*(7–8), 1292–1301.

Lv, J., Gu, F., Zhang, W., & Guo, J. (2019). Life cycle assessment and life cycle costing of sanitary ware manufacturing: A case study in China. *Journal of Cleaner Production, 238*, 117938.

Mangla, S. K., Luthra, S., Jakhar, S., Gandhi, S., Muduli, K., & Kumar, A. (2020). A step to clean energy-sustainability in energy system management in an emerging economy context. *Journal of Cleaner Production, 242*, 118462.

Meyers, P. G. (1964). *U.S. Patent No. 3,137,184*. Washington, DC: U.S. Patent and Trademark Office.

Narayanan, D., & Jagadeesha, T. (2020). Process Capability Improvement Using Internally Cooled Cutting Tool Insert in Cryogenic Machining of Super Duplex Stainless Steel 2507. In *Innovative product design and intelligent manufacturing systems* (pp. 323–330). Singapore: Springer.

Neves, A., Godina, R., Azevedo, S. G., & Matias, J. C. (2019). A comprehensive review of industrial symbiosis. *Journal of Cleaner Production, 247*, 119113.

Nouari, M., List, G., Girot, F., & Coupard, D. (2003). Experimental analysis and optimization of tool wear in dry machining of aluminium alloys. *Wear, 255*(7–12), 1359–1368.

Pandey, H., Garg, D., & Luthra, S. (2018). Identification and ranking of enablers of green lean Six Sigma implementation using AHP. *International Journal of Productivity and Quality Management, 23*(2), 187–217.

Patala, S., Salmi, A., & Bocken, N. (2020). Intermediation dilemmas in facilitated industrial symbiosis. *Journal of Cleaner Production, 261*, 121093.

Rusinko, C. (2007). Green manufacturing: An evaluation of environmentally sustainable manufacturing practices and their impact on competitive outcomes. *IEEE Transactions on Engineering Management, 54*(3), 445–454.

Sharma, A. K., Tiwari, A. K., & Dixit, A. R. (2016). Effects of Minimum Quantity Lubrication (MQL) in machining processes using conventional and nanofluid based cutting fluids: A comprehensive review. *Journal of Cleaner Production, 127*, 1–18.

Sharma, J., & Sidhu, B. S. (2014). Investigation of effects of dry and near dry machining on AISI D2 steel using vegetable oil. *Journal of Cleaner Production, 66*, 619–623.

Siegel, R., Antony, J., Garza-Reyes, J. A., Cherrafi, A., & Lameijer, B. (2019). Integrated green lean approach and sustainability for SMEs: From literature review to a conceptual framework. *Journal of Cleaner Production, 240*, 118205.

Simboli, A., Taddeo, R., Raggi, A., & Morgante, A. (2020). Structure and Relationships of Existing Networks in View of the Potential Industrial Symbiosis Development. In *Industrial symbiosis for the circular economy* (pp. 57–71). Cham: Springer.

Singh, M., Rathi, R., Khanduja, D., Phull, G. S., & Kaswan, M. S. (2020). Six Sigma Methodology and Implementation in Indian Context: A Review-Based Study. In *Advances in intelligent manufacturing* (pp. 1–16). Singapore: Springer.

Sony, M., & Naik, S. (2020). Green lean six sigma implementation framework: A case of reducing graphite and dust pollution. *International Journal of Sustainable Engineering*, *13*(3), 184–193.

Sreejith, P. S., & Ngoi, B. K. A. (2000). Dry machining: Machining of the future. *Journal of Materials Processing Technology*, *101*(1–3), 287–291.

Sun, Y., Bi, K., & Yin, S. (2020). Measuring and integrating risk management into green innovation practices for green manufacturing under the global value chain. *Sustainability*, *12*(2), 545.

Toppan Environmental performance data (2019) www.toppan.com/en/sustainability/environment/

Wang, Z. Y., & Rajurkar, K. P. (2000). Cryogenic machining of hard-to-cut materials. *Wear*, *239*(2), 168–175.

5 Circular Economy
Assessing a Progress of Resources Efficient Practices in Hotel Industry

Mudita Sinha and Leena N. Fukey
School of Business and Management, CHRIST
(Deemed to be University), Bangalore, Karnataka, India

CONTENTS

5.1 INTRODUCTION

Industries have battled to maintain an equilibrium in a value chain context amongst their ecological impacts, community welfare and cost benefits in recent years. This compels the industries to employ circular economy (CE) to optimize resources and administer carbon emissions (Urbinati *et al.*, 2017; Winans *et al.*, 2017). Industries need to revolutionize the need for resource depletion rates for some new economic model – CE facilitates the creation of an efficient and regenerative resource model by optimizing the resource used and the waste generated (Guo *et al.*, 2017; Mangla *et al.*, 2018). Things are changing at a greater speed than ever, the challenges of today are being converted into profitable opportunities of tomorrow. CE is one such hidden opportunity seeking to be utilized. CE also adds to the economy of both industry and nation by creating investment and new job opportunities, optimizing the cost of materials, stabilizing product prices, improving the resilience of the supply chain and reducing environmental impacts (Lieder and Rashid, 2016). In the context of production activities and business process, the preposition to enhance the sustainability of the value chain has become a contemporary issue (Jose Chiappetta Jabbour and Beatriz Lopes DeSousa Jabbour, 2014; Alcalde-Heras *et al.*, 2018; Brown and Bajada, 2018; Mishra *et al.*, 2018).

CE is a recent credo that refers to business sustainability through innovative production and consumption models (de Sousa Jabbour *et al.*, 2018). CE refers to the production of goods and services in an economy with limited intake of raw materials as well as eliminated/reduced waste generation in the process. In other words, it is taking a break from the linear economy to create a sustainable cycle. Circular tourism can be articulated as a model to create a fruitful circle to produce services without wasting the precious resources of the planet (Girard and Nocca, 2017).

CE is a sub-structure of the sustainable development model. The values governing CE are sustainable procurement, responsible consumption and extended recycling treatment. It works on three primary principles: first being natural capital conservation and development by managing the limited stocks and balancing the production of renewable resources; second principle aims at optimizing resource production by products, components and materials that are always of the highest utility quality, both for technological and biological cycles and the last principle speaks about promoting the efficiency of the system through the identification and design of negative externalities such as water, air, soil and noise pollution, climate change or resource-related harm to health (World Economic Forum, 2016). Though not a new concept, with the enhanced environmental issues, CE as a field is gaining more attention. Human desire to fulfil not just the needs but the luxuries has been gone far beyond the environmental tolerance, the threshold has been exhausted. A transition to a sustainable socio-ethical society has become the pressing priority, adoption of circular economic principles seems to be the way out. Benefits of adopting CE include environmental advantage, by directly reducing the effects of gas emissions, landfills, etc., economic advantage by saving money due to reutilization and social advantage by enabling people to join hands and reduce the social impacts like climate change.

Tourism, a powerful but increasingly complicated contributor to the economy in recent years, is one of the fields where preservation of the environment and the conservation of resources are significant. Hotels are an important component of the tourism industry. They play a vital role in generating revenue for the sector. This study focuses on the circular economies of the hospitality industry and feasible methods adopted to follow principles to reap benefits. Average waste generation rate of a 5-star property is around 2.28 kg/guest/day (Pham Phu *et al.*, 2018) and if single occupancy is considered for a 50-room hotel the water production rate is 114 kg and an average of 16760.28 kg for the world's largest hotel – The First World Hotel, Malaysia with an inventory of 7,351 rooms. Almost 83% of the hotel-generated waste is recyclable (Geissdoerfer *et al.*, 2017). The waste generated can thus be utilized again saving money, instead hotels only recycle an extremely limited amount of waste generated and leave away the rest for landfills. With accelerating importance given to sustainable development, the marketing of a Green Hotel has become goal of the hotel industry. The hotels thus must cultivate green consciousness and strengthen their green management to achieve the joint enhancement of the people and the organization. The CE proposes to reduce the impact on the environment while stimulating the economy through the development of businesses and new revenue streams (Kalmykova *et al.*, 2018). CE can contribute to sustainability through conditional, beneficial or trade-off relationships as a system that minimizes waste generation and emissions and mitigates material and energy loops to preserve sources (Geissdoerfer *et al.*, 2017). Management professionals have widely recognized CE practices and related activities; however, its methodological research evaluation is still unexplored (Korhonen *et al.*, 2018).

5.2 LITERATURE REVIEW

The environmental degradation effected the modern society we live in today. From manufacturing sector to the service sector, all of them majorly follow the linear economy model, resulting in mass output of waste from their operations (Figure 5.1).

Sustainable development can be addressed by the concept of CE that is continuously gaining grip and is progressively seen as a wide-ranging or partial solution to these challenges (Geissdoerfer *et al.*, 2017). According to Larsson (2018), CE is a commercial arrangement in which manufacture and distribution are controlled to use and re-use the same reserves repeatedly reducing the waste production (Figure 5.2).

As per Nocca (2017), CE principle of repair, recycle and reuse originated way back by Boulding (1966). CE is connected to modified consumption patterns and

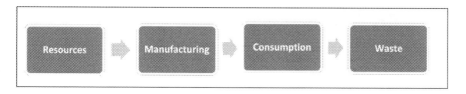

FIGURE 5.1   Linear economy model.

Source: Author's interpretation.

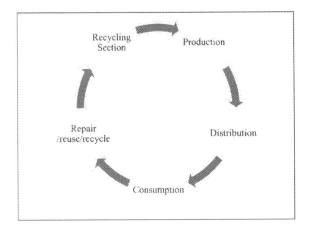

FIGURE 5.2 Circular economy model.

Source: Author's interpretation.

improved business operations, i.e. rather than having ownership of products to shifting to renting, revamping, repairing, reconstructing/renewing and recycling (Figure 5.3), which continuously emphasize its role in improvement of societal and economic importance. Kjaer (2015) visions about future lifestyle choices which looks very triggering and are currently need of the hour (Table 5.1).

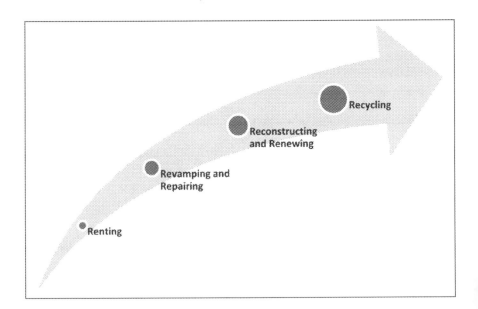

FIGURE 5.3 Upcycling modes in circular business.

Source: Author's interpretation.

TABLE 5.1
Tomorrow's Lifestyle Choice

Twentieth Century	Twenty-First Century
Hyper consumption	Meaningful consumption
Ego	Community
Meconomics	Weconomics
Ownership and credit	Access and sharing
Social status	Reputation
Sustainable	Unsustainable
Capital P = Profit	Capital P = Purpose

Source: Kjaer, 2015, p. 168.

Manniche *et al.* (2017a, 2017b) symbolized green economy to 'cradle to grave' and CE to 'cradle-to cradle' and differentiated implementation of CE in two phases: CE 1.0 focuses about circular fundamentals and novelties that can be applied at the organization level in the immediate future, whereas CE 2.0 articulates a systemic evolution and modification for a novel socio- and techno-economic model. Ming (2006) said that while the notion of the CE has been employed in the agriculture and construction industry in China, its introduction and implementation into tourism is extremely limited. Pattanaro and Gente (2017) agreed to same and advocated that the current research indicates limited deliberation on CE practices. The service sector in this scenario draws lesser attention but it is the silent destroyers (Álvarez *et al.*, 2001). The significant chunk of the service sector are the hotels. The CE practices adopted by a hotel are largely dependent on the age, size, operations management techniques, stakeholder environmental pressures and chain affiliations. Though hotels and other accommodation units actively participate in sustainability edges but the primary focus is on commercial aspect to gain competitive advantages that is the reason the hotels focus more on reduction of energy and water consumption and waste generated, whereas the environmental and socio-cultural aspects are often neglected (Girard and Nocca, 2017).

The facility's age is related to the environmental practices adopted by the hotel. Findings prove that the modern establishments house a greater number of good environmental practices as compared to the age-old properties. This is due to the knowledge available to the modern hoteliers. Understanding and sensitivity of the stake holders towards the ecological concerns is imperative for the hotel firms (Martínez-Martínez *et al.*, 2019). There are five categories of stakeholders responsible for adoption of CE practices namely – workforce, customers, shareholders, the government bodies and NGOs (Govindan and Hasanagic, 2018). The practices can be easily adopted and worked towards new properties but in case of the old properties, many are not built to adopt these practices since the knowledge about the sustainable practices seem to less of a need at their given time frame. The size of the hotel is directly proportional to the environmental practices adopted by the chain. Larger establishments are under the stakeholders' pressure to do so (Pham Phu *et al.*, 2018). Since the administration is centralized and funds exist to initiate new and better practices, larger establishments seem to adopt the sustainable practices as compared to their smaller counterparts. The relationship with the hotel chains affiliation and adoption of CE, standardized processes are established by the hotel chains which enables to follow the SOPs set efficiently. Many of the chains follow a decentralized system for operations ranging from purchasing, to taking reservations, to laundry, etc., This system has proven to reduce the environmental impacts as compared to the non-affiliated properties. Systemic transition of CE will denote that the hotel sector will be considered as a set of circular flows of interconnected, more or less closed materials, permitting a cascade demonstration of the resources between activities or services like accommodation, restaurants, well-being and leisure, etc. (Manniche et al.,2017a, 2017b).

The operational techniques adopted by the hotels, namely process and product standardizations, time and motion studies, quality circles, act as factors assisting the reduction of environmental impacts (Kim, 2012), benefitting both the organization and the environment. Thus, better operation techniques enable greater sustainable practices.

Oberoi Udai Vilas of Udaipur, Rajasthan, India is an eco-friendly property under the parent organization – Oberoi Hotels and Resorts. The resort has adopted extensively evident sustainability practices. The property has 87 rooms and began operations in the year 2002. Resembling a conventional Indian Palace with enclosures, well-kept gardens, mirroring pools and fountains incorporated with modern day sustainability systems like the solar power usage, rain water harvesting, waste water treatment, etc., making the property eco-friendly and beautiful for the guests to visit. The complete property is spread across 20 hectares of land of which 9 hectares comprises of gardens, 8 hectares comprises the wildlife reserves and only 3 hectares of land comprising the resort for accommodation and services. This is their first step towards being eco-friendly, that is, by restricting their revenue generating area to 3 hectares and avoiding over development in terms of construction and human intervention. Central Zoo Authority of India has a tie-up with property for intention of the wellbeing of the wildlife reserves and a dedicated hotel staff for feeding and protecting animals along with a veterinarian for the good health of the animals.

The Oberoi group have outdone themselves by following all the unmandated rules ethically and have stepped up to protect the heritage and natural environment at site. Building has incorporated traditional building methods. The temperature maintenance is done naturally through the mixture of indoors and outdoors, open air corridors, natural channels for rainwater harvesting. The material used in the construction are green marble, sandstone and wood sourced locally. The windows are double glazed and lime wash is used as a substitute for paint. Though the lime glaze needs to be done regularly and looks like a cost center, it is indeed not as the lime wash is 70% cheaper than paints. The local purchasing of consumable goods, water harvesting and better practices for energy usage makes the Oberoi Udai Vilas one of its kind in India.

What a company does or has done in terms of environmental issues can describe its contribution to the local habitat. A few examples of environmental practices that a company may engage in include are preparing an environmental plan, a written document with a detailed description of its environmental plan, communication of the environmental plan to shareholders or stakeholders, communciation of the environmental plan to employees, incorporation of an environment, health and safety (EHS) unit, and formation of a board or management committee dealing dedicatedly with environmental issues (Henriques and Sadorsky, 1999). An interesting finding from the study done by Álvarez et al., (2001) is that stakeholders of an organization are proactive in pushing the management to adopt sustainable practices when a third-party strong force like the government regulations are made mandatory. This is because the stakeholders view ahead for their organizations to exist for an indefinite long time and following the set rules seem to be the first step for the continuity. Important environmental impacts are caused by tourism which in turn generates pressure on local supplies creating negative externalities. Not only land but water and food produce large amount of waste. Tourism causes road bottlenecks, air and sound pollution along with emission of carbon dioxide. (Rico et al., 2019). Managing environment has become an important step for tourism to maintain the economical position (Blanco et al., 2009). By incorporating steps of environment conservation tourism sector can attract clients who are inclined towards environment protection and can also make it as a unique selling preposition for promoting hotels as a competitive advantage (Alonso-Almeida et al., 2015; Llach et al., 2013; Perramon et al., 2014) and also improving its image (Jang et al., 2011; Perramon et al., 2014).

Resource consumption in hotels and other accommodations has been in focus in many studies. Environmental reports of hotel companies have included some consumption indicators. However, the influence of various operational characteristics on energy and water consumption is still not fully researched that has further scope of improvement.

Physical and operational parameters influence the performance. Physical parameters include key dimensions, structure details, design type, geographic and climatic location, age, energy and water systems, operation and maintenance details, energy and water resources also available locally, regulations and costs or slabs on the use of energy and water. The use of resources in hotels is further influenced by a great deal specific operating characteristic. These include several operations schemes for underground utilities including refreshment points, laundries, swimming pools and spas, leisure and commercial centers, etc., services offered, fluctuations in employment levels, changes in customer preferences regarding internal comfort, as well as culture and awareness of resource consumption. Several environmental benchmarking the schemes have been developed internationally, some of which they gained publicity and market penetration.

5.3 METHOD

This research article sketches the prominence of sustainability practices in the hotel industry and provides insights to the CE practices embedded in the business planning and operations of the selected hotel chains. This research is exploratory in nature and embraces multiple business case analysis for specific selected hotel chains. This method is incredibly useful in understanding the different aspects of sustainability in hotels. The research focus is primarily on physical parameters.

The evaluation of the specific instances provided some of numerous and heterogeneous strategic situations related to organizational preparations of inter-firm collaborations with respect to sustainability. This research also recognizes few potential objectives for scholarly research and professional contributions.

5.4   HOTELS AND CIRCULAR ECONOMY – CASE FILES

Sustainability is believed to be the ability to cater to the needs of the present without depleting the future resources. A discussion on the topic is the need of the hour and is crucial for the future development of luxury and hospitality means. At the core of the issue, it boils down to what we define sustainability and to what extent we are willing to go to take ownership over our practices.

The study entails a compilation of the sustainability practices of four international hospitality companies and the effectiveness of their practices, namely, The Accor Group, The Indian Hotels Company Limited, The Hilton and the Shangri-La hotel group. We have focused on their sustainability practices and ventures to help promote sustainability amongst all their stakeholders.

The intent of this study was not only to deepen our knowledge on sustainability, but also educate ourselves on how to responsibly bring about a change in the industry.

5.4.1   I – INDIAN HOTELS COMPANY LIMITED

The Indian Hotels Company Limited and its subsidiaries offer a combination of Indian neighborliness and world-class administration. They operate under the name of Taj, SeleQtions, Vivanta and Ginger.

The Tata group was founded by Jamsetji Tata in 1868 and is now a global enterprise that comprises over 100 independently operating companies. It operates in six continents with the motive of providing good quality of services to the people. Through long-term stakeholder value creation based on 'Leadership with Trust', the Tata brand stands as a lasting promise behind its businesses, many of which are industry leaders.

The Taj group aims at building a sustainable environment by reducing the impact they have on the environment and using sustainable practices, conserving resources and reusing and recycling key resources.

5.4.1.1   Sustainability Practices and Conservation of Environment at Taj Group of Hotels

Apart from the common goals that the organization intends to achieve, most of the brands now a days are now initiating practices in order to conserve the environment and ensure a sustainable environment. Being sustainable refers to conserving the resources for the future generations. The company aims at judicious use resources so that the resources don't get depleted and can be used for the upcoming generations. As a part of protecting the environment, the company develops environmental sustainability through efficient management of all assets and resources. In addition to this the brand has also introduced EARTH (Environment Awareness & Renewal at Taj Hotels), which help in integrating their operations and product design and development to help in functioning in tandem with the environment. To improve the aesthetics with regards to the environment theme, the brand has also introduced Green EARTH Rooms at some properties that feature small but significant changes that are eco-friendly and resonate with guests.

5.4.1.1.1 Water

Water efficiency in Taj mostly comprises upgrading to water-efficient equipment and having administrative controls in place. Recycling increased equipment efficiency, and rainwater harvesting could minimize freshwater withdrawal and preserve bodies for communities. In key cities, they carried out water security assessments in hotels and calculated risks and strengthened our response to an issue.

In the year 2017–2018, withdrawal of water for the year factored in two more new components and they were bottled water and water from rainwater harvesting as these were not included in the calculations of the previous years. Water recycled in 2017–2018 refers only to recycled STP (Sewage Treatment Plant) water. The consequence is an increased water withdrawal and decrease in recycled water and hence the years between 2017 and 2018 cannot be compared to later years. Some properties use onsite bore wells. Via rainwater harvesting and recharge, TAJ protects onsite groundwater and by treating the output and sewage, they make sure groundwater does not get contaminated.

Sustainability is an important aspect regardless of the location or business. The team at TAJ properties put in a lot of effort to maintain the efficient and judicious use of water and electricity. They try to make sure that they provide good quality guest experience whilst cutting down the carbon footprint. TAJ uses locally sourced environmentally friendly material while constructing areas near natural parks, forests, and fragile coastlines. IHCL is one of those hotel groups who participates and engages themselves in external environmentally friendly activities such as turtle conservation and coral reef restoration.

All of the above-mentioned efforts have resulted in 58,000 tons of CO_2 emissions avoided and more than 3.2 million KL of water reused and recycled. Over the last 3 years, we have conserved enough water to fill 338 Olympic sized swimming pools. We Achieved this as 44 of our hotels have become 'zero water discharge' hotels, meaning that not a drop of water is discharged into the municipal sewers.

5.4.1.1.2 Energy

Taj has been implementing green initiatives since the early 1990s under the Eco Taj policy. Taj hotels and resorts launched EARTH in 2006, a movement that works to minimize the impact of its businesses on the environment (Table 5.2). Under this, they train employees to be environment friendly and conduct energy audits every three years. In 2016–2017, 60 Taj hotels were awarded EarthCheck Gold Certification, 16 were awarded Silver Certification and in the year 2018–2019, 8 hotels were platinum-certified, 63 were gold certified, 8 were silver certified and 2 were bronze certified.

TABLE 5.2
Green Initiatives

Property name	Location	Initiatives
Taj Coromandel	Chennai	Generates electricity from windmills.
Taj West End	Bengaluru	Uses solar water heating systems and has saved around 51 thousand liters of fuel over 3 years.
Vivanta by Taj	Coimbatore Jaipur	Uses wind energy to power 80% of their power requirements. (August 2019 – By Khushboo Goyal)
Taj Rambagh Palace – It is used to source power through the Rajasthan state electricity board, where more than 55% of the power was produced by warm force stations.		Introduced a 2.1 MW wind turbine generator at a breeze ranch at Kaladongar in Jaisalmer, which presently supplies 70% of the yearly power necessities of the hotel. (Indian Hotels Company Limited Corporate Sustainability Report 2012–2013, p. 39)
Vivanta by Taj – Holiday village – While maintaining its gardens, they generate some garden waste and spend 0.1 million rupees to dispose of it.	Goa	They convert the waste into char by thermal decomposition and make fuel blocks that are used in tandoori fire grills.

Source: Authors Interpretation

According to the latest ITP (Inspection Test Plan) report, the tourism sector accounts for up to 1% of global emissions and to try and reduce this, Taj hotels have implemented various energy conservation initiatives like switching to LED and CFL lights, installing Variable Frequency Drives (VFDs) on high power motors like condensers, cooling tower fans and air handling units. The CFL lights consume 78% less energy.

5.4.1.2   Energy and Climate Change

The properties primarily generate energy from

a. Direct sources like diesel, LPG, petrol, fuel oil and charcoal. Direct renewable sources include solar, biogas and biomass.
b. Indirect sources – electricity grids. Renewable sources are agro power and hydropower.

In 2012–2013, Taj had 22 hotels that produce electricity from solar energy, though it was not a major source of electricity it still shows the sustainability measures that the hotel takes. Both direct and indirect sources of renewable energy contribute to 6.77% of the total energy consumption. Because of the joint efforts on renewable energy, Taj increased the total energy by 92.9% (Indian hotels company limited corporate sustainability report 2012–2013, p. 37).

5.4.1.3 Results of the Initiatives

- In the year 2016–2017, the Taj group was able to reduce 55,603,013 kg of CO_2 emissions, which is the same as taking 18,659 cars off the road. (UNITED NATIONS GLOBAL COMPACT COMMUNICATION ON PROGRESS Year: 2016–2017).
- In the year 2017–2018, 58,580 tonnes of CO_2 emissions were avoided by switching to renewable energy.
- In the year 2016–2017, 13% of their energy mix consisted of renewable energy, which has helped them cut down 65189.2 kg of CO_2 out of their total greenhouse gas emissions.
- In the year 2017–2018, 21% of their energy mix was renewable energy. 252677658 MJ of energy was procured from renewable sources. This resulted in a cut of 58.58 KT of CO_2 emissions further reducing their carbon footprint.
- By using alternate fuels like Bio gasoline, piped natural gas (PNG), compressed natural gas (CNG), they have reduced their scope 1 emissions. Scope 1 emissions are direct emissions caused by the activities of the organization or any activity that is under their control.
- 60 Taj properties met 21.75% of its energy needs through renewable energy sources.

5.4.1.4 Limited Resources with Circular Economy Concepts

The move towards a CE framework can possibly change the cordiality business. CE for IHCL is 'restorative and regenerative by plan', rather than a straight 'take, make and arrange'; economy. As a mass purchaser of different items, our waste age is intently attached to the structure of our graceful chain and the accessibility of elective materials. Successful administration of natural waste and limiting plastics are our center territories. We have done certain developments with the waste produced in mass from our Hotels like the creation of Biodiesel from Waste Cooking Oil. Advancement and dispersion of new advances are irreplaceable for Business Sustainability. One such advancement venture for changing over waste oil into biodiesel that can control vehicles have been perceived at Tata Innovista in 2019. This venture whenever received by additional in the neighborliness business will affect the social biological system by making more employments in the capable reusing division.

South Asia's biggest cordiality organization has propelled a 'Waste to Innovation Challenge' over its 149 lodgings on the side of Earth Hour. As a major aspect of the test, the inns will contend with one another to concoct inventive plans to change over waste created inside the inns into significant items. The most significant and maintainable activity will be decided for expected scaling over the organization.

Taj MG Road, Bengaluru is among the leaders in the business to change over waste oil into usable biodiesel, which is being utilized to control cross breed transports. Moreover, the side-effect of glycerin is utilized to make cleansers for the network; subsequently, empowering giving a coordinated methodology towards round economy.

Dr PV Murthy, Executive Vice-President and Global Head – Human Resources, IHCL said 'IHCL has been taking an interest in Earth Hour for a long time with extraordinary eagerness'. The current year's drive suitably mirrors our responsibility towards ecological stewardship by incorporating supportable practices as a feature of our everyday tasks through inventive thoughts. We are additionally pleased to declare that inside a year, we have wiped out 2,000,000 plastic straws from our biological system since our promise to the reason.

Furthermore, all IHCL lodgings would watch a one hour switch off of power from 8.30 to 9.30 pm on 30th March, 2019 combined with a large group of intriguing visitor and worker commitment exercises, for example, star looking, narrating, meals in obscurity and pedal for the Planet.

5.4.1.4.1 Improving Biodiversity

We all know that plastics have created a huge crisis in the world, especially in the hotel industry where there is a lot of plastic usage seen. It's affecting the marine food chains. The IHCL has come up with new aspects to reduce the usage of plastics in the hotel. They replaced all the plastic straws with bamboo straws, which reduced the usage of around 2 million plastic straws. Taj exotica resort claims to be the first resort for zero use of plastic, it's situated in Andaman with 45 acres of forest and mangroves in Havelock islands, and supports green tourism, which protects the eco systems and also the marine animals. The resorts have completely replaced plastic bottles with glass bottles. IHCL has come up with a mission for reducing single use plastics and moving on to recyclable plastics. This also creates awareness of the eco systems and its importance to the guests as well.

Taj exotica resorts and spas mainly focus on sustainable luxury and promoting green tourism. They do sustainability practices like rainwater harvesting. Water bottling plants and also, they claim that not a single tree has been affected by their construction of the property. The resort has also partnered with pollution control board to keep the island plastic free, and collection of recyclable plastics in the mainland. The resort mainly focuses on changing the human behavior towards usage of plastics and conserving the forests or the marines. All these practices have contributed a lot for conserving wildlife and improving biodiversity.

5.4.1.4.2 Promoting Biodiversity

Taj group has a luxury lodge which is named as Pashan Garh. A 200 acre wildlife with stone houses provides luxury wildlife travel experience. Here the main motto of theirs is to provide luxury experience by preserving the forest and biodiversity conservation. The hotels/resort is located in the outskirts of some of the famous tiger reserves. The taj safari at panna national parks has naturalists who give the awareness of wildlife conservation and biodiversity. They take the guests through the process of conserving forests and its importance. The guests are experienced with spotting different species of bird in the forest; the naturalists provide efficient knowledge about the conservation of forests. They have not disturbed any of the wildlife species or the forest while constructing the rooms. In the nearby villages they also conduct awareness classes about the wildlife and the importance of biodiversity. The villagers' existence is only because of the forests and every day is an earth day for them (Amit Kumar, lodge manager, taj safaris). 'Our job is to not only create awareness about the wildlife but also concentrate on conservation [Dipu Sasi, naturalists, Taj safaris]'. IHCL has part-nered with the pardha youth, which is a tribal youth's group near panna national park. They have an initiative saying walking with pardha where they showcase their livelihood which takes tourism to the next level. The project also concen-trates on the promotion of natural habitat and wildlife conservation. The natural-ists have trained the pardha women to showcase their own traditional cooking styles which enables them to earn and as a well. And this project has helped over 20 families so far.

5.4.2   II – Shangri-La

Shangri-La Hotels and Resorts is the trading entity of Shangri-La International Hotel Management Limited, a Hong Kong-based multinational hospitality company.

Shangri-La Group has brought together all CSR activities under the umbrella brand Sustainability inside five key regions: Environment, Health and Safety, Employees, Supply Chain and Stakeholder Relations.

5.4.2.1   Water

Shangri-La Group is one of the world's premier developers, owners and operators of hotel properties. It currently owns over 100 hotels globally in 75 destinations under the Shangri-La, Kerry, Hotel Jen and Traders brands. Their Environmental Management System (EMS) gives a hearty structure to arranging, overseeing and controlling environmental impact.

Water is a scarce resource in many places around the world. Water holds a signifi-cant portion in hotel utility bills. Shangri la keeps the responsibility not to consume more than necessary. They strive to moderate their effect on the accessibility of freshwater in environmentally sensitive areas.

In 2017, Shangri la was working with 'WATERISLIFE', which is a globally recognized charity, to provide safe and clean drinking water to underprivileged communities.

In November 2017, they installed the water treatment systems in three elementary schools in remote communities in west southern China.

Shangri la recognizes responsible water management is an important issue for the planet, thus in 2018 they completed tests which were new water-saving shower heads that will reduce the consumption of water by a certain percentage without compro-mising the showering experience.

In the 2019, sustainability report, an aggregate of 18,056 super liters (ML) of water was devoured by properties, including for the most part freshwater from metropolitan supplies for drinking, cooking, cleaning, water system and recre-ational civilities. Recognizing the importance of water management there were actions taken.

For example – Their resorts in Boracay, Cebu, the Maldives, Mauritius and Yangon work desalination plants utilizing groundwater or seawater to deliver fresh-water for utilization.

Also, 53 hotels roll out a shower head replacement program in staff facilities, which could reduce water consumption by 20–30% by limiting the rate of flow, and this helped the company to achieve estimated water savings of 120 CBM per year.

WATER STRESS: It occurs when the demand of water exceeds the available amount during a certain period of time. Therefore to reduce the burden, Shangri la utilizes the Aqueduct Water Risk Atlas created by the World Resources Institute (WRI) to distinguish which of their properties are situated in zones encountering 'High' or 'Amazingly High' levels of water pressure which identifies the area which has 'High' or 'Extremely High' levels of water stress.

In Shangri La Bangalore, there are water treatment plants which supply treated and recycled water to cooling towers which is then used for centralized air conditioning and additionally used to water the garden outside and inside the hotel as well.

5.4.2.2   Energy
Cutting Edge Technology used in Shangri la hotel:

1. Energy management control system (EMS)
2. Room management control system
3. LED lamps
4. New generation environmentally friendly chiller with VSD.

The Shangri-La Hotel selected all these technologies as part of a pilot project to improve their carbon footprint

It is always important to select the right contractor and equipment for the job. The competitive tendering process is desirable to keep the project price down.

5.4.2.3   Chiller Upgrade
The main features of upgrading the chiller were:

1. Very low load with good efficiency.
2. Chiller includes 'Rapid Re-start' after failure, 43 seconds after locating power source

Management took many initiatives for power saving and a greener environment. They tried to change many appliances and divided the work in two phases

Guest Room Energy Management System

Occupancy sensors are an effective way to save energy. They work with smart control. As the guest adjusts the key parameters (e.g. Temperature, Flow, etc.) the system recognizes and repeats it as long as the room is occupied. Needless to mention that when guest moves out the occupancy sensor sets the parameters to energy saver mode.

5.4.2.3.1   Renewable Resources
About 80% of the hotel's power supply can be generated from solar and wind energy, which helps to attenuate commercial energy consumption. Also, in fact, in line with Forbes Magazine, '100 square meters of solar panels help keep a 100-watt light bulb

running for the entire year with 8 days, 8 hours and 14 seconds of energy required.' Installation of solar panels reduces quandary energy consumption not only that it also reduces the expenditures on liquefied petroleum gas, up to 30%.

Shangri-La as a brand has taken various measures all across the globe in its properties to make sure a sustainable environment, the Shangri-La Hotel, Bengaluru is additionally specializing in power saving and seeing green initiatives within the hotel that employment to attenuate environmental impact and conserve biodiversity and manage waste, water and energy.

Taking the instance of Thailand within which as could be a suitable country for solar power, and the way the government has implemented policies to assist develop solar power, endorsing its environmental and financial benefits. The business class Shangri-La Hotel has installed solar panels across a 938 m^2 space on the rooftop to power a solar water heating. Installation of solar panels reduces quandary energy consumption not only that it also reduces the expenditures on liquefied petroleum gas, up to 30%.

Thailand's Integrated Energy Blueprint Plan was launched in the beginning of 2015, which brought out the ambitious thrust on renewable energy. Further, the plan had the Energy Efficiency Plan (EEP) for community with a target of recovering 30% of the country's total energy consumption from renewable sources.

Not only this, Shangri-La had Renewable Watch organized its Eleventh Annual Solar Energy Conference in India. The conference companies likewise as specialized experts within the sector, among other things. The event provides a hunt of emerging trends and opportunities within the sector, the impact of recent policy and regulatory initiatives, discussing risks and challenges, and showcasing the most recent innovations.

5.4.2.3.2   Limited Resources

Sustainability is basically focusing on meeting needs of the present by not compromising on the future. Sustainable practices are very important as it is the key for a greener earth. Following sustainable practices is very important in organizations also.

Shangri-La in Delhi is promoting breathe easy stay, which is basically growing plants which can aid natural air purification. The hotel has also placed an electronic air purifier in order to keep the environment free from pollution. Snake plants that are placed in hotel lobbies have purification properties, which remove formaldehyde and nitrogen oxides. Peas Lilly, aloe Vera, money plants are also placed in different places at the hotel, which also helps in improving air quality.

Other sustainable practices followed in Shangri-La are water aerators, which helps to reduce the wastage of water. Installation of charcoal tandoor by replacing PNG-based tandoor is another example of sustainable practices. Even the cleaning techniques using chemicals are replaced with steam clean.

Whichever material that is non-sustainable can cause pollution. So, it is very important to dispose of everything in the right manner. One plastic bag which is disposed of in the wrong manner can cause pollution. It also affects the Marine life and it takes around 100 years for a plastic bag to decompose. Shangri-La promotes the usage of cloth bags and took the initiative to clean the surrounding areas and promote a plastic-free environment. They also conducted cloth bags painting competitions in order to promote the usage of reusable bags.

They also try to include their sustainable practices in the house keeping department also. They use the products like soaps, shampoo and other amenities from Ming fai who produces natural products and has a bio-degradable packaging and soy ink printing. They even use responsible washing and cleaning products that are mild to launder the linen, which maintains the quality of the linen. Most of the Shangri-La hotels around the world have implemented green housekeeping and low temperature laundry practices. These sustainable practices can make changes in the environment and can help achieve a greener earth. Each step taken by an organization to promote sustainable practices can influence a lot of people and helps to make a difference.

5.4.2.3.3 Improving Biodiversity

Shangri-La group has been committed to address how their daily operations affect climate change overall. As a big group of hotels, they have their social responsibility to care about how their operation affects climate change and improve sustainable practices to reduce their overall carbon footprint. Shangri-La Group also registered as a member of United Nations Global Compact in 2011 and implemented their 10-principle operational structure to provide a framework for sustainable practices in the hospitality industry. Shangri-La has also been committed to the protection of endangered species in order to save it for the future generations.

The Shangri-La group has the group's green policies also promotes resource efficient facilities and environmentally responsible SOPs to ensure a sustainable environment of operations. They have also implemented the SOPs to use lesser and lesser amounts of single use plastics, thereby reducing their overall plastic footprint. Shangri-La has also introduced water bottling plants to adapt more sustainable practices across the entire chain of hotels. Shangri-La hotel group has also innovated low temperature laundry facilities, composting, smart technology food waste management and plantings of microgreens and herbs in their gardens.

Shangri-La's Care for Nature Project also promotes the conservation, safeguarding and restoration of biodiversity. Shangri-La is located in some of the most beautiful and ecologically diverse areas of the world and have taken the responsibility to look after the local biodiversity in these pristine locations for future generations, with partnership along with local governments, NGOs and local communities around their properties to protect and safeguard the flora and fauna in the local environment.

Shangri-La has also been very prominent in voicing its concern against single use plastic, especially stirrers and straws made of plastic in the food and beverage sector of their chain of hotels and resorts.

5.4.3   III – Hilton

Hilton Worldwide Holdings Inc. is an American multinational hospitality company that manages and franchises a broad portfolio of hotels and resorts. They are known for their exceptional quality in service and mindfulness in their business activities.

5.4.3.1 Water

Hilton Hotels is making an active attempt to conserve water in their operations (Table 5.3). They do this because they understand the need for quality water and its availability in the future. Their hotels have many guidelines that need to be followed to reduce their water consumption.

They made a long-term goal to reduce their water consumption in their operations by 50% by the year 2030. In the year 2012, Hilton achieved its short-term goal of a 10% reduced usage. This achievement was possible by consistently using less water by 10.2% for last 4 years.

In the time period of 2011–2012, the water use fell from 25.45 million gallons to 24.72 million gallons. The graph for reduction in water consumption between 2009 and 2012 is given as follows.

Unfortunately, Hilton has failed itself and its customers because they are now behind target. They have reduced their water consumption by only 23% in all hotels and 22% in their managed hotels. It has taken them over a decade to reduce water consumption by 22% and with only 10 more years to go and less than halfway there, Hilton is going to have a tough time catching up. The following graph is a representation of their water consumption over the years till 2019.

Hilton partnered itself with the World Wildlife Fund (WWF) in 2015 for sustainable water use and seafood. The WWF has a history of partnering with hotels to support the sustainability efforts. Hilton currently uses the WWF's Water Risk Filter to provide data for the water conservation efforts in all their hotels. Hilton hotels have adopted sustainable housekeeping products and water-efficient showers. These measures help reduce 30% water consumption. It indirectly resulted in better efficiency and less usage of toxic chemicals employed for cleaning. Coupled with sustainable laundry technology, the consumption is down by 45%.

Over 1,700 of their hotels, franchised and owned, have installed smart/drip irrigation. They save 6,000 gallons every day by recycling greywater at Hilton Garden Inn Dubai. Hilton properties even follow rules on low-flow fixtures.

Increasingly exposed to water stress. A 40% shortfall of water is forecasted by 2030. Preserving water also includes preserving aquatic ecosystems and not over fishing. Hilton pledged to do this in June 2016, on World Oceans Day. By 2022, they aim to ban the procurement of endangered species from all their properties, supply their kitchens with at least 25% of their ingredients from Marine Stewardship Council

TABLE 5.3

Water Consumption

Year	Gallons of Water Per Square Foot
2009	78.8
2011	76.0
2012	73.8

Source: Authors Interpretation.

certified and Aquaculture Stewardship Council certified sources and the remaining must be actively working towards these certification and sustainable improvements. Hilton Singapore actually became the first hotel in Asia to serve sustainable seafood and became the first to serve MSC certified fish at 41 properties across Europe.

5.4.3.2 Energy

Hilton is a leading hospitality company, with 14 world-class brands having more than 5,400 properties in 106 countries. It is dedicated to fulfilling its mission to be the world's most hospitable company. The company includes Hotels and Resorts. With the intention of keeping the low spending on energy (the 2nd highest operating cost), low-cost, high-impact steps and a portfolio-wide certification to ISO 50001 was persuaded, resulting in low consumption of energy. According to its recent report, it has saved nearly a quarter of a billion dollars from energy efficiency projects over the past 4 years. In recent years, the major topic of concern is how an organization can be sustainable and ecofriendly to the maximum. The hospitality industry is said to be the most carbon-producing industry in the world. The carbon footprint is a major concern for climate change and is a major issue in the world right now. Hilton as a brand has been trying for the past years to be sustainable and rely on renewable energy as much as possible. Now to solve this problem, Hilton has come up with a product known as 'LightStay' in 2009, in fact, they received Environment Leader's 2016 Product of the Year Award. They were awarded this for their innovative approach to sustainability. LightStay enables the brand to manage and measure the waste, water, and energy across all its hotels in all the countries. The company could achieve this due to 'Living Sustainably', by its Travel with Purpose program, that sets sustainability goals to reduce energy consumption, carbon, waste and water. Hilton has exceeded its five-year goal to reduce water consumption by 10% and waste output by 20% one and two years ahead of schedule, respectively. With its help Hilton has been able to reduce the carbon emissions equivalent to 390,350 cars, they have also been able to save up to $1 Billion as utility costs.

Hilton is the only hotel company featured on the prestigious Fortune Change the World list for both 2018 and 2019. With the help of this product, it became the first global Hospitality company to achieve ISO 50001, one of the first multinational companies to certify its entire system globally and achieve one of the largest-ever certifications of commercial buildings. It's also the first brand to certify commercial buildings to the US Department of Energy's Superior Energy Performance program in the world. Hilton has been able to reduce its energy use by carbon output by 20.9% and 14.5%, waste output by 27.6% and water use by 14.1% in six years after LightStay was invented, resulting in an estimated $550 million of cumulative savings. LightStay uses prehistoric data and uses it to estimate the future usage of energy, the cost and the impact on the environment.

5.4.3.2.1 Social Impact

Hilton Effect Grants are installed and awarded annually with the sole purpose of helping advance Travel with Purpose – 2030 Goals; it is committed to double investment in social impact and cut environmental footprint by half – this they will do with responsible hospitality in the entire value chain.

Responsible sourcing by supporting local entrepreneurs and fostering responsible practices in the supply chain.

Creating opportunities by providing skills and employment to the local community, investing in innovative local solutions, and deploying commercial engines for local hotel owners to grow.

Enabling business by accelerating access to tourism's economic benefits for local businesses.

Respecting destinations by preserving local cultural and natural heritage and fostering respect for all.

Preserving resources by tracking against science-based targets for carbon, driving water stewardship, re-engineering waste and innovating every day for sustainable operations.

5.4.3.2.2 Improving Biodiversity

HILTON, being a global hospitality company, Hiltons and resorts is a global brand of full services and the flagship brand of American multinational hospitality company Hilton and the original company was founded by Conrad Hilton.

Every march for past five years Hilton (different properties around the world) has been taking part in WWF's earth hour, in 2015 Hilton announced its commitment with WWF for collaborating on extending sustainable seafood efforts globally and advancing Hilton food wastage efforts, Hilton worldwide views sustainability as key to run and establish its business it believes that every company should contribute back to the nature. Hilton has also taken steps in order to cut off environmental footprint as its social impact investment.

Food waste: as Hilton has taken an effort to reduce its environmental footprint as its social impact investment and half to be done by 2030 so food production is the major cause for deforestation, water extraction, habitat and biodiversity loss and yet one third of food wastage takes place with regard to the waste audits done by the hotel it is proved that

Almost 41% of waste produced by the hotel is food waste, Hilton is trying to focus on measurements capabilities to better understand the food wastage level in present time. Around 890 hotels have globally implemented the composting programs. Hilton has also started food donation programs to reduce the food wastage.

Single use plastic: As in today's world plastic has become a major issue as it has become a threat to the communities and environment thus as a globally known brand and as a origination Hilton has observed its need to understand this issue and take few steps to overcome this problem thus the hotel tries to focus on reusable items where ever possible. From 2019, the brand has stopped using disposable plastic.

Biodiversity conservation: Guests are encouraged and provided with information regarding protected areas and native wildlife and plant species found in the areas within proprieties influence zone, which is provided at the concierge desk as well as hotel members are trained to provide this information to the guests.

5.4.4   IV – Accor

Accor group is a worldwide hotel operator with more than 5,000 properties, 740,000 rooms in more than 110 countries and 300,000 human workforces.

Its various hotel brands are categorized into Luxury brands, Premium brands, Midscale brands, Economy brands, Work spaces and Play spaces.

5.4.4.1   Sustainable Practices to Support Remote Location Tourism

One of their major projects for sustainability is the Planet 21 Program, where they have identified various sustainability goals with respect to the Sustainable Development Programme for 2030 and the Sustainable Development Goals (SDGs) validated by the United Nations General Assembly.

Planet 21 focuses on two specific issues: healthy and sustainable food with ban on food wastage; a move towards carbon-neutral buildings.

The first example would be funding the restoration of the Chapelle Rouge at Karnak, Egypt (A historical place).

The next one would be Raffles Maldives Meradhoo resorts focusing on sustainable seafood from the local sourced market.

The next would be the idea of the Mantis Hotels, an eco-escapes and lifestyle hotel, deeply rooted to African Conservation developed with a vision of sustainability and a world where humans and animals could co-exist. It develops tourism to remote locations with sustainability in it. It includes eco-lodges, boutique hotels, game reserves, ski lodges, chalets and yachts.

The main objective of Mantis is to be committed to conservation, restoration and each property is sensitive to its surroundings in respect of the buildings, environment and local community.

Mantis has also partnered with EcoPlanet Bamboo to utilize bamboo charcoal air purifiers in all its hotels and also using it for filtration of water thereby reducing use of plastic water bottles in remote locations. Also, EcoPlanet Bamboo helps the hotel in acquiring luxury textiles from bamboo fiber. The next would be Movenpick hotels and resorts, which is a Green Globe certified hotel, helps in sustainable development of the local community, the area where the hotel is located.

NOTE: All the hotel brands mentioned here fall under Accor Group of Hotels. The hotel brands operates most or few of its hotels/resorts in remote location, which helps in development of remote location tourism sustainably

5.4.4.1.1 Water

Hotels consume large quantities of water for day-to-day usage, which further adds on to the importance of water sustainability to be ingrained into a hotel's policy to ensure that all corporate measures are being taken to protect one of the planet's most endangered resources.

Accor is a French multinational company that owns, operates and franchises hotels, resorts and vacation properties. They are Europe's largest hospitality company and the sixth largest worldwide. They operate in 100 countries with 4,800 hotels and 280,000 employees. They have started a program named 'Planet 21' that aims to showcase their commitment to sustainability for 2020. Their objectives are structured around four pillars: working with their employees, involving their clients, innovating with their partners and involving local communities.

Accor claims to be the pioneer in setting up a comprehensive environmental policy. In 1994, they set up an environmental department exclusively with the role of creating an environmental business approach, which now consists of 56 correspondents worldwide led by an environmental manager.

Accor began a program in 2006, titled the 'Earth Guest' program, that aimed to reduce water and energy consumption and was divided into two areas: a social responsibility development area (EGO) and environmental responsibility development area (ECO).

The Accor policy of 2007 has eight key projects. The Hotelier's Environment Charter project laid down the guidelines and clear instructions for local management to save resources. It also states how guests should be approached regarding environmental issues. In a separate project aimed towards Water and Energy conservation. Importance of water efficiency was highlighted by Accor's proactive approach. Accor set up solar panels in their hotels to sustainably produce hot water for guest usage.

The various programs, research and initiatives by Accor truly showcase their commitment to water sustainability and sustainable development as a whole which is an important factor for hotels to consider in today's world.

5.4.4.1.2 Energy

Present in over 90 countries that hosts 500,000 guests everyday providing them with water, energy and food and expect their guests to look at the world differently in a more engaged and responsible way. Their vision drives the change towards positive hospitality globally. Planet 21 is an initiative by ACCOR group of hotels where they endeavor to engage their guests in an even more sustainable hotel experience, they act as an inclusive company for their employees while diversity being one of their assets, they also co-innovate with their partners, they work hand in hand with local communities for a positive impact.

Planet 21 – 2011–2015 achievements

In the year 2015, which was the last year of their first season of PLANET 21, they managed to tighten CSR governance, which resulted in the speedup of their PLANET 21 objectives.

Planet 21 – Present day scenario

Accor Hotels owns over 30% of its hotels. In order to embark hotels on the energy transition path, Accor is building/designing more and more carbon-neutral buildings, as well as renovating existing ones with the aim of increasing energy efficiency.

The following are Accor group's commitments towards PLANET 21

a. Make sure all the renovated and upcoming hotels are low-carbon buildings
b. Reduce energy consumption by 5

The below mentioned are a few Hotels under Accor that are heading towards energy transition:

a. Tile roofs (made of insulating materials that reflect the sun's rays)
b. Double glazing
c. Energy efficient air conditioning
d. Heat recovery system
e. Presence sensors to switch corridor lights on and off
f. 530 m^2 of solar panels that covers almost 45% of hot water requirements in the hotel

Limited resources with regards to sustainability has always been a challenge, but it can be done, it is how you make the best out of the natural resources and give back to nature. Accor has been moving towards sustainability for quite a while now, their PLANET21 drive has been very successful.

To bring into light issues about soil degradation and deforestation, our forests have so many resources that are a boon for many industries. Accor hotels have been using wood and from sustainably managed forests to make eco-friendly beds. Bedspreads and pillowcases from recycled bottles. They have been very successful in terms of marketing their ECO-DESIGN beds and other products, which they use in many of their properties.

Accor has been doing its part in terms of having a drive for planting trees. Their PLANET21 development program Accor initiated a towel reuse and reforestation project – they have planted over 3 million trees worldwide. Also, to be more sustainable they have many measures for unproductive soil, combating erosion and carbon sequestration, motivating its suppliers to move into agroforestry. Now in many hotels for cleaning purposes of the pools, they use a lot of chemicals. Novotel München city Arnulfpark has been using flora and fauna to clean their pools.

Accor hotels have been trying to reduce carbon footprint for a long time now and they have been successful for having measures like using LED LIGHTS, flow regulators and thermal insulation of complex objects, precise configuration of the systems so that making it possible to reduce carbon footprint and have more efficient energy systems.

Accor hotels have also adopted the innovative pillows and duvets filled with bedding fibers which are made from recycled PET.

Accor also has been practicing certain methods such as collecting and reusing cooking oil, because of the rising issues of food waste because the hospitality industry generates so much, most of the hotels use them as fuel and fuel also used for airplanes.

The Accor group has stated that construction and renovation waste constitute about 70% of the total waste produced by the hotel. Hotels have been taking measures to refurbish the wastes so that recycling is used, waste such as light bulbs, complimentary products and food have generated more waste so how efficiently Accor uses it has made a huge impact on the environment.

5.4.4.1.3 Renewable Resources

The Accor group in 2007 with the help Agency for Environment and Energy Management had started their journey in sustainability in the area of solar energy. They initially started off with 100 hotels and slowly had a goal of equipping 200 hotels with solar panels by 2010. After realizing the effect and feasibility of solar energy, it had worked on reducing and implementing waste management and thus also informing other customers and hotel brands on their step in sustainability.

This was also started alongside the commitment with ADEME in enhancing customer and employee awareness of environmental concerns and also to create awareness among the general public about the advantages of renewable energy. Since the beginning of their agreement back in 2003, Accor and ADEME have been working on reducing the impact hotels have on the environment.

5.5 DISCUSSION AND ANALYSIS

As we discussed and reviewed the cases to understand the sustainability practices, it is understood that hotel organizations do have sustainable practices in place unfortunately some are only on papers and not really practiced. Perhaps, there are significant evidence of sustainable practices followed by many and some can be referred and practiced by other hotel organizations. Brand-wise good practices with respect to sustainability are discussed in this section. Water efficiency in Taj mostly comprises upgrading to water-efficient equipment and having administrative controls in place. Recycling increased equipment efficiency, and rainwater harvesting could minimize freshwater withdrawal and preserve bodies for communities. They have introduced heat pumps and condensation water circulation pumps also. While building any new property Taj has started giving a lot of importance towards building the property using sustainable structure and contributing towards the society and ecological development. In the year 2016–2017, the Taj group was able to reduce 55,603,013 kg of CO_2 emissions which is the same as taking 18,659 cars off the road. In the year 2017–2018, 21% of their energy mix was renewable energy. 252,677,658 MJ of energy was procured from renewable sources. By using alternate fuels like Bio gasoline, PNG, CNG, they have reduced their scope 1 emissions. Scope 1 emissions are direct emissions caused by the activities of the organization or any activity that is under their control. They have done certain developments with the waste produced in mass from our Hotels like the creation of Biodiesel from Waste Cooking Oil. Taj Group has played a very crucial role in launching several of India's key tourist destinations, working in close association with Indian government. Vivanta Dal View, located in Srinagar, Jammu and Kashmir is one of the best examples of remote tourism. The IHCL (the mother company of Taj) has come up with new aspects to reduce the usage of plastics in the hotel. Taj exotica resort claims to be the first resort for zero use of plastic. Taj group has a luxury lodge which is named as Pashan Garh. Here the main motto of theirs is to provide luxury experience by preserving the forest and biodiversity conservation. The project also concentrates on the promotion of natural habitat and wildlife conservation.

Shangri-La started conserving water by using water saving shower heads and by collaborating with different charities. The shower head replacement program in staff facilities which can reduce water consumption by 20–30% by limiting the rate of flow was implemented, and this helped the company to achieve estimated water savings of 120 CBM per year. The Shangri-La Hotel selected all these technologies as part of a pilot project to improve their carbon footprint – Energy management control system (EMS), Room management control system,

LED lamps and New generation environmentally friendly chiller with VSD. They also conducted a chiller upgrade. SISC (Shangri-La Institute of Sustainable Communities) addresses the three fundamentals: natural reserves, community learning centers, and willing communities, to successful ecotourism that stretches across three areas of Shangri-La: the Napahai area (Kesong), Bazhu village and Baima Xueshan Nature Reserve. Not only that, a foundation of an 'ecotourism resource center' has been laid in Shangri-La's Dukezong Old Town, in order to help connect all of the existing sites. Shangri-La in Delhi is promoting breathe easy stay, which is basically growing plants which can aid natural air purification. The hotel has also placed an electronic air purifier in order to keep the environment free from pollution. About 80% of the hotel's power supply can be generated from solar and wind energy, which helps to attenuate commercial energy consumption. Shangri la not only offers care and hospitality to its guests but also to the local communities where the property operates.

Hilton hotel groups have been trying to conserve water for more than a decade now and they have only been successful in reducing 22% water consumption. They have partnered with WWF and even pledged to preserve their aquatic ecosystems and even not to overfish. Hilton is also making international efforts to understand the local nature of water risk and its need to be sustainable locally. The remote location tourism provided by Hilton is the Hilton Shillim Estate Retreat is Located in the Sahyadri mountain range within the 320-acre Shillim Estate, which is a very remote location near Lonavala. Hilton also improves its biodiversity conservation by reducing single use plastics as well as their food wastage.

Accor group of hotels have created strategies and projects for increasing sustainability. One of the projects are the Planet 21 Program where they have identified various sustainability goals with respect to the Sustainable Development Programme for 2030 and the SDGs validated by the United Nations General Assembly. Planet 21 focuses on two specific issues: healthy and sustainable food with ban on food wastage; a move towards carbon-neutral buildings. Through this program they reversed their usual increase in energy consumption to a decrease of 7.2% in energy consumption. Accor began a program in 2006, titled the 'Earth Guest' program, that aimed to reduce water and energy consumption. Accor has been moving towards sustainability for quite a while now, their PLANET21 drive has been very successful. They have also been successful in reducing their carbon footprint. They are planning to make sure that all the renovated and upcoming hotels are low carbon buildings and to reduce the energy consumption by at least 5%. The Sofitel The Palm Dubai and The Novotel München City Arnulfpark are examples for hotels that have reduced their carbon footprint and increased their energy efficiency. The Accor group in 2007 with help from the Agency for Environment and Energy Management had started their journey in sustainability in the area of solar energy. They initially started off with 100 hotels and slowly had a goal of equipping 200 hotels with solar panels by 2010. They also partnered with ADEME to increase awareness on sustainable energy.

5.6   CONCLUSION AND SUGGESTION

Resorts and hotels are often central to a community, so they have the ability to raise awareness and change behavior. Companies must come up with solutions rather than problems, especially when it comes to sustainability. This study was to analyze the sustainability features of 4 main hotel groups Taj, Shangri-La, Accor and Hilton. The data was collected on seven main areas of the four hotel groups. The areas were water, structural, energy, remote location tourism, limited resources, improving biodiversity and solar. According to our study, all the four hotels follow practices and policies that are praiseworthy, but in some cases, these hotels could use another look at their practices. When it comes to water management, Shangri-La follows an innovative practice of showerhead replacement and reusing their water for their cooling systems which makes them the leader in water management practices. This also shows their innovation in structural design to be able to integrate their water management system and cooling systems to their architecture initiates an effective system. In terms of energy management, Hilton leads the way in introducing their 'light stays' model to drastically cut down on energy consumption. The strategy they used was to meticulously set targets and keep these goals on priority. Another noteworthy initiative is the IHCL who diversify their energy sources giving priority to renewable resources. Environment conservation is a very important factor to look into, which all hotels equally play their part in keeping in check their carbon emissions and waste disposal methods. The IHCL takes this initiative to the next level by introducing their presence in Pashan Garh by turning their property into a wildlife conservation territory. With respect to social impact, by far the Accor hotels have performed excellently by involving the owners, employees, clients and the local communities into their PLANET21 Sustainability project and analyzing the impact of their decisions on their entire environment. Thereby showing that all the above hotels need to be open to adapting to current wants and future requirements.

The main take away from this study is that companies must benchmark their own standards of sustainability and aim to improve them instead of aiming to compete with other groups. Guests are more conscious and aware of their impact the key here is to try to understand the effect of one change on other processes. By integrating their practices, they would likely see a combined result and would be able to sustain in the long run. This research can work as guidelines for the hotel chains to equate the CE practices implemented by other chains. This research can also provide policy makers to focus on exploring new practices of CE to support the environment.

The use of resources in hotels is further influenced by a great deal by specific operating characteristics. The research was limited to physical parameters and excluded detailed study and influence of operating characteristics. The question remains, however, how valid and reliable they are (benchmarks) and how they are established, needs further research.

REFERENCES

Alcalde-Heras, H., Iturrioz-Landart, C., & Aragon-Amonarriz, C. (2018). SME ambidexterity during economic recessions: The role of managerial external capabilities, *Management Decision*, Available at: https://doi.org/10.1108/MD-03-2016-0170

Alonso-Almeida, M. M., Bagur-Femenias, L., Llach, J., & Perramon, J. (2015). Sustainability in small tourist businesses: the link between initiatives and performance. http://dx.doi.org/10.1080/13683500.2015.1066764.

Álvarez Gil, M., Burgos Jiménez, J., & Céspedes Lorente, J. (2001). An analysis of environmental management, organizational context and performance of Spanish hotels. *Omega*, 29(6), 457–471. doi:10.1016/s0305-0483(01)00033-0

Blanco, E., Rey-Maquieira, J., & Lozano, J. (2009). Economic incentives for tourism firms to undertake voluntary environmental management. *Tourism Management*, 30(1), 112e122. http://dx.doi.org/10.1016/j.tourman.2008.04.007.

Boulding, K. E. (1966). The Economics of the Coming Spaceship Earth. In J. Henry (Ed.), *Environmental quality issues in a growing economy* (pp.3–14). Baltimore, MD: Johns Hopkins Press.

Brown, P. J., & Bajada, C. (2018). An economic model of circular supply network dynamics: Toward an understanding of performance measurement in the context of multiple stakeholders. *Business Strategy and the Environment*, 27(5), 643–655.

de Sousa Jabbour, A. B. L., Jabbour, C. J. C., Godinho Filho, M., & Roubaud, D. (2018). Industry 4.0 and the circular economy: A proposed research agenda and original roadmap for sustainable operations. *Annals of Operations Research*, 270(1–2), 1–14.

Geissdoerfer, M., Savaget, P., Bocken, N. M., & Hultink, E. J. (2017). The circular economy – A new sustainability paradigm? *Journal of Cleaner Production*, 143, 757–768.

Girard, L. F., & Nocca, F. (2017). From linear to circular tourism. *Aestimum*, 70, 51–74.

Govindan, K., & Hasanagic, M. (2018). A systematic review on drivers, barriers, and practices towards circular economy: A supply chain perspective. *International Journal of Production Research*, 56(1–2), 278–311.

Guo, B., Geng, Y., Ren, J., Zhu, L., Liu, Y., & Sterr, T. (2017). Comparative assessment of circular economy development in China's four megacities: the case of Beijing, Chongqing, Shanghai and Urumqi. *Journal of Cleaner Production*, 162, 234–246.

Henriques, I., & Sadorsky, P. (1999). The relationship between environmental commitment and managerial perceptions of stakeholder importance. *Academy of Management Journal*, 42(1), 87–99. doi:10.2307/256876

Jang, Y. J., Kimb, W. G., & Bonn, M. A. (2011). Generation Y consumers' selection attributes and behavioral intentions concerning green restaurants. *International Journal of Hospitality Management*, 30(4), 803e811. http://dx.doi.org/10.1016/j.ijhm.2010.12.012.

Jose Chiappetta Jabbour, C., & Beatriz Lopes DeSousa Jabbour, A. (2014), Low-carbon operations and production: Putting training in perspective. *Industrial and Commercial Training*, 46(6), 327–331.

Kalmykova, Y., Sadagopan, M., & Rosado, L. (2018). Circular economy – From review of theories and practices to development of implementation tools. *Resources, Conservation and Recycling*, 135, 190–201.

Kim, J. (2012). International Journal of Hospitality & Tourism Administration, 13, 195–214, p. 208. Retrieved from https://www.scientific.net/AMM.291-294.1478" https://www.scientific.net/AMM.291-294.1478

Kim, Y. J., Palakurthi, R., & Hancer, M. (2012). The environmentally friendly programs in hotels and customers' intention to stay: An online survey approach. *International journal of hospitality & tourism administration*, 13(3), 195–214.

Kjaer, A. L. (2015). Understanding tomorrow's consumer landscape. In R. Talwar (Series Curator and Editor), Wells, S., Koury, A., & Rizzoli, A. (Eds.), *The Future of Business* (pp. 163–170). Fast Future Publishing.

Kjaer, A. L. (2015). Understanding tomorrow's consumer landscape. *The Future of Business, Fast Future Publishing, Tonbridge,* 163–8.

Korhonen, J., Nuur, C., Feldmann, A., & Birkie, S. (2018). Circular economy as an essentially contested concept. *Journal of Cleaner Production, 175,* pp. 544–552. doi: 10.1016/j.jclepro.2017.12.111

Larsson, M. (2018), *Circular business models: Developing a sustainable future.* Cham: Palgrave Macmillan.

Lieder, M., & Rashid, A. (2016). Towards circular economy implementation: A comprehensive review in context of manufacturing industry. *Journal of Cleaner Production, 115,* 36–51.

Llach, J., Perramon, J., Alonso-Almeida, M., & Bagur-Femenías, L. (2013). Joint impact of quality and environmental practices on firm performance in small service businesses: An empirical study of restaurants. *Journal of Cleaner Production, 44,* 96–104. doi: 10.1016/j.jclepro.2012.10.046

Mangla, S. K., Luthra, S., Mishra, N., Singh, A., Rana, N. P., Dora, M., & Dwivedi, Y. (2018). Barriers to effective circular supply chain management in a developing country context. *Production Planning & Control,* 29(6), 551–569.

Manniche, J., Larsen, K. T., Broegaard, R. B., & Holland, E. (2017a). Destination: A circular tourism economy. A handbook for transitioning toward a circular economy within the tourism and hospitality sectors in the South Baltic region. Nexoe: Centre for Regional & Tourism Research. Available at: https://crt.dk/media/90318/Cirtoinno-handbook_CRT_05102017-002-.pdf *(Accessed 8 June 2018).*

Manniche, J., Topsø Larsen, K., Brandt Broegaard, R., & Holland, E. (2017b). Destination: A circular tourism economy: A handbook for transitioning toward a circular economy within the tourism and hospitality sectors in the South Baltic Region. Project Mac-CIRTOINNO. Nexø: Centre for Regional & Tourism Research (CRT).

Martínez-Martínez, A., Cegarra-Navarro, J.-G., & García-Pérez, A. (2019). Knowledge agents as drivers of environmental sustainability and business performance in the hospitality sector. *Tourism Management, 70,* 381–389.

Ming, Q.-Z. (2006). The new development conception of tourism circular economy and its systematic operation mode. Available at: http://en.cnki.com.cn/Article_en/CJFDTOTAL-YNSF200605010. htm *(Accessed 7 April 2018).*

Mishra, S., Singh, S. P., Johansen, J., Cheng, Y., & Farooq, S. (2018). Evaluating indicators for international manufacturing network under circulareconomy. *Management Decision,* available at: https://doi.org/10.1108/MD-05-2018-0565

Nocca, F. (2017). The role of cultural heritage in sustainable development: Multidimensional indicators as decision-making tool. *Sustainability,* 9(10), 1882+.

Official Website – First World Hotel Plaza, Malaysia. Retrieved from https://www.guestreservations.com/first-world-hotel/booking?msclkid=a11057046a5118f80acd0963e21c9f78

Official Website – Marriott Group. Retrieved from https://www.marriott.com/default.mi

Official Website – Shangri-La Group. Retrieved from https://www.shangri-la.com/?WT.mc_id=SLIM_201703_GLOBAL_SEM_YAHOO_SLIM-Brand_Brand-Exact_shangri%20la_EN&gclid=CMiKwNf7m-gCFYycjgodv74BgQ&gclsrc=ds

Official Website – Taj Group. Retrieved from https://www.tajhotels.com/

Pattanaro, G., & Gente, V. (2017). Circular economy and new ways of doing business in the tourism sector. *European Journal of Service Management,* 21(1), 45–50.

Perramon, J., Alonso-Almeida, M. M., Llach, J., Bagur-Femenias, L. (2014). Green practices in restaurants: Impact on firm performance. *Operations Management Research,* 7(1e2), 2e12. http://dx.doi.org/10.1007/s12063-014-0084-y.

Pham Phu, S., Hoang, M., Fujiwara, T. (2018). Analyzing solid waste management practices for the hotel industry. *Global Journal of Environmental Science and Management*, 4(1), 19–30. doi: 10.22034/gjesm.2018.04.01.003

Rico, A., Martínez-Blanco, J., Montlleó, M., Rodríguez, G., Tavares, N., Arias, A., & Oliver-Solà, J. (2019). Carbon footprint of tourism in Barcelona. *Tourism Management*, 70, 491–504.

Urbinati, A., Chiaroni, D., & Chiesa, V. (2017). Towards a new taxonomy of circular economy business models. *Journal of Cleaner Production*, 168, 487–498.

Watson, R., Wilson, H. N., Smart, P., & Macdonald, E. K. (2018). Harnessing difference: A capability-based framework for stakeholder engagement in environmental innovation. *Journal of Product Innovation Management*, 35(2), 254–279.

West, J., Salter, A., Vanhaverbeke, W., & Chesbrough, H. (2014). Open innovation: The next decade. *Research Policy*, 43(5), 805–811.

Winans, K., Kendall, A., & Deng, H. (2017). The history and current applications of the circular economy concept. *Renewable and Sustainable Energy Reviews*, 68, 825–833.

Witjes, S., & Lozano, R. (2016). Towards a more circular economy: Proposing a framework linking sustainable public procurement and sustainable business models. *Resources, Conservation and Recycling*, 112, 37–44.

World Economic Forum, 2016. Intelligent Assets Unlocking the Circular Economy Potential (Online). Available at: (*Accessed 4 April 2018*).

World Travel & Tourism Council (WTTC) (2019). Annual Report. Available online:https://www.wttc.org/about/media-centre/press-releases/press-releases/2019/travel-tourism-continues-strong-growth-above-global-gdp/

WEB REFERENCES

https://s3-us-west-2.amazonaws.com/ungcproduction/attachments/cop_2019/477876/original/IHCL-SustainabilityReport'19.pdf?1566968536

http://www.greenhotelier.org/best-practice-sub/case-studies/benchmarking-with-earthcheck-taj-hotels-group/

https://www.ihcltata.com/press-room

https://www.forbes.com/sites/rahimkanani/2014/05/04/taj-hotels-resorts-and-palaces-on-fighting-poverty-and-sustainability/#519dd3185662

https://books.google.co.in/books?id=aIkcDQAAQBAJ&pg=PA19&lpg=PA19&dq=sustainability+in+structural+design+of+taj+hotels&source=bl&ots=09ceAh-fxn&sig=ACfU3U1WPqkWn-

https://www.flamingotravels.co.in/blog/2016/07/the-taj-rambagh-palace-the-residence-of-maharajah-of-jaipur

https://www.unglobalcompact.org/participation/report/cop/create-and-submit/active/374561

https://s3-us-west-2.amazonaws.com/ungc-production/attachments/cop_2019/477876/original/IHCL-Sustainability-Report'19.pdf?1566968536

https://www.news18.com/news/india/indian-hotels-initiate-green-practices-342176.html

https://www.greenhotelier.org/best-practice-sub/case-studies/benchmarking-with-earthcheck-taj-hotels-group/

https://renewablewatch.in/2019/08/09/green-stays/#:~:text=In%202017%7D18%2C%20Indian%20Hotels,solar%20and%20wind%20energy%20footprint

https://s3-us-west-2.amazonaws.com/ungc-production/attachments/cop_2019/477876/original/IHCL-Sustainability-Report'19.pdf?1566968536

https://s3-us-west-2.amazonaws.com/ungc-production/attachments/cop_2019/477876/original/IHCL-Sustainability-Report'19.pdf?1566968536

https://www.tajhotels.com/en-in/taj/pashan-garh-panna-national-park/rooms-and-suites/

https://www.shangri-la.com/group/our-story/community-and-social-impact
https://www.shangri-la-sustainability.com/about-us
https://www.environmentalleader.com/2013/10/hilton-sustainability-report-company-beats-water-goal/
https://thegate.boardingarea.com/earth-day-2017-and-what-hilton-is-doing-to-help-save-the-environment/
https://cr.hilton.com/wp-content/uploads/2020/04/Hilton-Water-Fact-Sheet.pdf
https://sitecore-cd.shangri-la.com/-/media/Project/Shangri-La-Group/OurStory/Files/Community-And Social-Impact/Sustainability-Reports/2019-Sustainability-Report_ENG.pdf
https://digitalcollections.sit.edu/cgi/viewcontent.cgi?referer=https://www.google.com/&https redir=1&article=2350&context=isp_collection
https://www.greendesignconsulting.com/post/2018/06/05/solar-energy-a-great-opportunity-for-thailand
https://thecsrjournal.in/how-shangri-la-hotels-are-going-green-in-india/
https://renewablewatch.in/2018/07/03/highlights-solar-power-india-2018-conference/
https://indiacsr.in/shangri-la-on-a-mission-to-promote-sustainable-environment/
https://newsroom.hilton.com/brand-communications/news/lightstay-a-decade-of-managing-our-environmental-and-social-impact
https://www.environmentalleader.com/products/hilton-worldwide-lightstay/
https://sustainablebrands.com/read/marketing-and-comms/hilton-worldwide-saves-250-million-from-energy-efficiency-in-4-years
https://www3.hilton.com/en/hotels/india/hilton-shillim-estate-retreat-and-spa-PNQSHHI/attractions/wellness-programs.html
https://cr.hilton.com/wp-content/uploads/2019/05/Hilton-Corporate- Responsibility-2018.pdf
https://press.accor.com/raffles-maldives-meradhoo-opens-overwater-villas/?lang=en
https://tophotel.news/accor-showcases-future-of-eco-travel-at-the-iconic-mantis-founders-lodge/
https://www.tourismtattler.com/articles/sustainable-tourism/eco-friendly-hotels-mantis-collection/67592
https://group.accor.com/en/Actualites/2017/02/15/tourism-at-the-service-of-sustainable-development
https://group.accor.com/en/commitment/positive-hospitality/acting-here
https://group.accor.com/-/media/Corporate/Group/PDF-for-pages/Brand-book/Accor-Brands-Book-2020.pdf
https://all.accor.com/gb/sustainable-development/index.shtml
https://www.greenhotelier.org/our-themes/policy-certification-business/summary-of-accors-planet-21-initiative/group.accor.com › media › 20180330-AH-DDR2017-EN

6 A Governance-Practice Framework for Remanufacturing in the Indian Automobile Sector

Shikha Verma
Production and Quantitative Methods, Indian Institute of Management, Ahmedabad, Gujarat, India

Anukriti Dixit
Public Systems Group, Indian Institute of Management, Ahmedabad, Gujarat, India

CONTENTS

6.1 INTRODUCTION

Sustainability initiatives in India fall under the umbrella of the commitments of the Millennium Development Goals charted out in the Millennium Summit of the United Nations. While a plethora of work has been done in the country for renewable energy, forest management, grassroots innovation, social engagement of communities as well as of businesses, there is a need to set a larger context for the environmental, social and governance aspects of a sustainable state. A circular economy framework is associated with converting the limiting aspects of a linear economy into a technology driven set of industries that can help drive sustainability through its valuable chain (Ellen Macarthur Foundation, 2016). This prospect is more potent for a country like India, where the manufacturing sector is relatively young and policies are adaptive according to the needs of the people. In this paper, we posit that it is possible to integrate governance and business initiative toward a circular framework along with driving changes in consumer mindsets. We therefore propose a three-tier approach toward demonstrating a circular framework through the introduction of a remanufacturing ancillary for India's automobile sector. In recent times, the investment in India's automobile sector has risen and the sector contributes to 22% of the Indian GDP and FDI (Make in India, 2015). A remanufacturing incentive of the industry backed by a nudge-based approach for altering consumer behaviour that will aid in promoting sustainability in the long run for this sector (Lehner *et al.*, 2016). Given the extractive and energy intensive nature of the automobile industry, an initiative to promote remanufacturing will require rigorous participation of all the actors and this will lead to long-term benefits for the economy.

This paper explores literature from three different streams, which are as follows:

1. Behavioural public policy design for incentivizing more sustainable consumer behaviours (Thaler & Sunstein, 2008; McKenzie-Mohr, 2011)
2. Industry limitations with respect to remanufacturing of automobiles in India and similar emerging economies (Govindan, Shankar, & Kannan, 2016; Khan *et al.*, 2019a)
3. Sustainable governance practices and policy interventions – best practices for automobile firms, remanufacturing supply chains and consumers (Zhang *et al.*, 2017; Mangla *et al.*, 2020)

Combining these sets of literatures, we generate insights for a governance-practice-based framework that can be utilized for future remanufacturing endeavours in the automobile industry. Through our three-pronged framework (Figure 6.1), we discuss three main actors in this scheme – the state, the automobile companies and the consumers. In addition, within the automobile firms, we include remanufacturing middle firms, resellers and informal markets structured around the dissembling and reprocessing of used vehicles. We begin by exploring all the three links but we would like to mention that even though the discussion on these links is separate, this does not imply their independence in real time. For a successful implementation of this conceptual framework, these links will have to function in synergy as amalgamated agencies and sub-agencies.

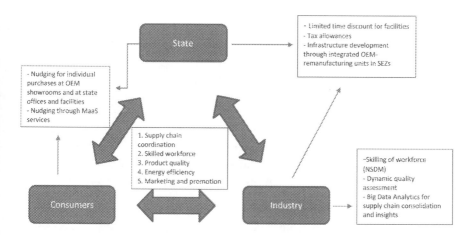

FIGURE 6.1 A governance-practice-oriented framework for a holistic approach for encouraging remanufacturing in the Indian automobile sector.

Source: A self-sourced schematic representation of common concerns and potential pathways to address those concerns for each of the actors.

6.2 LITERATURE REVIEW: PROBLEM AREAS AND OPPORTUNITIES FOR REMANUFACTURING IN THE AUTOMOBILES SECTOR

The automotive industry in India has historically given a large contribution in the country's GDP. However, in recent years, the Department of Heavy Industry (DHI), Government of India (GOI) has reported a marked recession in the automotive sector, which is driven largely by the following factors.

[.] slowdown in economic growth, high cost of vehicle finance, high interest rates, high fuel prices, high inflation and negative market sentiments, increase in the commodity prices, high customs duty on Alloy Steel, Aluminium Alloy and Secondary Aluminium Alloy, high rate of service tax and excise duty, high and varied rate of road taxes in the States, low growth of export markets, etc. Ministry of Heavy Industries and Public Enterprises has been consistently taking up the matter of providing some kind of stimulus package with prompt fiscal and other measures to put the industry back on track. As a result, in the interim budget for the year 2014–2015, reduction in excise duty in case of cars, two wheelers and truck chassis was announced. *Department of Heavy Industry, Ministry of Heavy Industries & Public Enterprises, Government of India website.*[1]

The measures of relaxing excise tax are one among the many measures that are proposed in the DHI's Automotive Mission Plan 2016–2026 (Automotive Mission Plan, 2016). This plan includes measures such as changes to trade policies, reduction in taxes as well as enhanced manufacturing infrastructure to mitigate the growing recession being faced by the industry. However, we find that with regards to end-of-life measures, emission control measures and environmental sustainability aspects of the industry, very little attention is paid, within the plan, to remanufacturing as a potential area of investment for policymakers.

With a resource intensive industry such as the automobile sector, a remanufacturing endeavour will not only result in reduction of emissions over the original manufacturing emission levels (Sutherland *et al.*, 2008), it will also guarantee
lower levels of landfill waste, lower costs and energy consumption levels (Govindan
et al., 2016). Remanufacturing in the automotive sector is also a high profitable
option with a profit margin as high as 20% (Mukherjee & Mondal, 2009) with
remanufacturing practices having the highest credit out of the overall remanufacturing practices (Kim *et al.*, 2008). In the United States, an organized remanufacturing
sector benefits from higher profit margins, high savings on costs as well as emissions and high energy savings from remanufactured engines (Amelia *et al.*, 2009).
It is noteworthy that this sector, in the US context, is largely industry driven as
opposed to being pushed by policymakers, as is seen in the context of European
Union (Mukherjee & Mondal, 2009).

However, in the Indian automobile context, despite an extended recession,
remanufacturing has remained excluded from the list of potential profitable, energy
and environment friendly options for the automotive industry. Khan *et al.* (2019b)
find that green ideology and its adoption in emerging economies sees insufficient
investment from both governance and industry partners with insufficient impetus on
dedicated infrastructure and policy frameworks for sustainable governance (Khan
et al., 2019b). Further, the reluctance for remanufacturing in the automotive sector
comes from several directions, including three primary issues:

- Industry reluctance due to supply chain issues, skilled workforce and quality hurdles and low awareness about profits: The supply chain for remanufacturing begins with collection, sorting, reconditioning, reassembling and
  marketing (Bras, 2014). However, in India, as compared to other emerging
  economies, there is little support for efficient design, production and quality
  testing (Rashid *et al.*, 2013; Chakraborty *et al.*, 2019). In addition, the entire
  supply chain described previously has several inputs from unorganized
  markets, with very few Original Equipment Manufacturers (OEMs) having take-back policies for their original auto parts (Govindan *et al.*, 2016).
  Similarly, lack of a skilled workforce is another barrier identified (Barker
  & King, 2007), along with an insufficiency of technologies to disassemble
  auto parts (Xian & Ming, 2011).
- Consumer mindset with regards to remanufactured equipment: There are
  various theories for consumers' behavioural change, including those of
  rational choice theories (Turaga *et al.*, 2010), community-based social marketing (McKenzie-Mohr, 2011), moral explanations of altruism vs. egoism
  (Prakash *et al.*, 2019). Among these, the choice architectural approaches
  offered by nudge theorists Thaler & Sunstein (2008) offer interesting ways
  to intervene through policies and through marketing into consumer choices.
  In our framework, we explore the different ways in which the nudge-based
  approaches can be employed for the push towards more responsible con-
  sumerism with respect to the automobile sector.
- Lack of affirmative public policies for the remanufacturing supply
  chain: The role of policy makers for the promotion of green energy and

technologies has been vastly signalled as a pivotal one (Khan *et al.*, 2019b; Mangla *et al.*, 2020). In the Chinese context, Zhang *et al.* (2017) suggest possible strategies for remanufacturing of automotive parts. Three of their primary recommendations include state mandated control of quality and stricter standardizations within the supply chain as well as extended responsibility for manufacturers and subsidization efforts for large scale remanufacturers. Further, Khan *et al.* (2019a) observe, in the context of Pakistan, similar barriers that can be addressed through more persistent and consolidated state action.

Our observations from the different sets of literatures lead to the understanding that while several studies propose government reforms (Govindan *et al.*, 2016; Zhang *et al.*, 2017; Khan *et al.*, 2019b) and suggest broadly that potential reforms· could be offered, an elaborate effort to chart out the state's role in intervening for a circular economy in the automobile remanufacturing sector, is exigent. This paper suggests a 'practice oriented policy framework' – one in which we delineate the role that state agencies and sub agencies could pay through collaborations, consolidation and streamlining of both the supply chain issues and the demand side (consumer behaviour) concerns. In addition, we discuss the implications for the automobile OEMs and the corresponding remanufacturing supply chains, along with strategies for better streamlining of the circular economy surrounding automobiles in India. Finally, we discuss the overall effectiveness and limitations of the recommendations offered in our framework.

6.3   EXPLORING THE FRAMEWORK

In a linear economy framework, the central solution for promoting green behaviour among manufacturers in the Indian context is through Corporate Social Responsibility (CSR). This includes a list of pro-social and pro-environmental activities, which can be promoted through a direct expenditure of the company's resources. CSR was made mandatory in India for firms who made over Rs 5 cr. net profit. However, as discussed by Mezzadri (2014), mandatory CSR represents an incompatibility with several of the manufacturing sectors and their goals. Thus causing 'proxy initiatives' to propagate, which are inauthentic in nature. In recent times, the recycling guidelines have also been revised. End of life vehicle recycling, while promoted in Japan and China has not been emphasized in India (Sakai *et al.*, 2014). The Ellen MacArthur foundation lays emphasis on the idea that sustainability can be re-conceptualized as a circular framework for manufacturers in India mainly because the manufacturing industry is relatively nascent and Green Manufacturing Supply Chain (GMSC) initiatives can be supplemented as well as bolstered with a circular framework. In the framework (Figure 6.1), we map the common concerns of the three actors in the centre of the figure and branching outwards from each actor, the proposed solutions have been summarized.

We begin by looking at the *state–industry link* and the *state–consumer link*. We then move on to exploring the bidirectional link between the automobile manufacturers and consumers as well as the auxiliary link between the industry and the informal or formal resellers and distributors, which includes all other middle-links and bridges within the automobile manufacturing and distribution network.

6.3.1 THE STATE-INDUSTRY LINK

Remanufacturing entails a process of disassembling, refurbishing, cleaning and remodelling the vehicle in such a way that it is fit for use as a new vehicle (Subramoniam *et al.*, 2009). A concept that has gained traction in the green supply chain literature is of Extended Producer Responsibility (EPR) means extending the producer's involvement design and production phases to collection, recovery and dismantling of used products and incentivizing them to take environmental considerations into account while designing products. However, in an imperfect legal system, small number of remanufacturers and 'one-time consumption' of consumers are barriers towards scalability in the production and adoption of remanufactured products.

Mandating remanufacturing up to a certain percentage is one tactic some governments have tried including that of China (Xiang & Ming, 2011). However, such measures can prove to be disastrous if the decision is perceived as a top down push. Such mandatory restrictions would, for instance, ensure that a percentage of the cars manufactured by an automotive company are remanufactured and put out for sale. However, in the absence of the demand for these cars or a deliberate marketing-based push by companies, on the brand-new car models, could lead to an oversight. The streets would be flooded with as many new cars as before and the old ones may find an alternative market albeit an insignificant one. There exists a danger of misconstruing circular frameworks to mean that the original rate of manufacturing remains the same while all the changes to the remanufacturing are brought with 'extending the customer base' as their motivation (Ellen MacArthur foundation, 2016). While extending the customer base may be one pecuniary externality of the market, once remanufacturing begins, it is not the goal of circular frameworks. The goal is to not only encourage conscious producers but also conscious consumers. Thus, in effect a framework designed to effect these changes will need to bear in mind both aspects. For promoting conscious producers, two major ideas we propose are strategic tax allowances and limited time discounts for facilities, which can be offered by the Indian government.

6.3.1.1 Strategic Tax Allowances

The tax concessions when offered for unlimited time periods and for a wide range of industrial activities run the risk of misuse. Since the implementation of the GST in India in 2017, it can be possible to tie together the promotion of several tax incentives due to the very nature of this tax. Since this tax requires input credits at every step (Vasanthagopal, 2011), remanufacturers can be offered a tax relaxation at specifically the expensive steps – such as spare part procurement – conditional to the item's usage for remanufacturing purposes. This relaxation may not even be a tax cut in particular but could also be in the form of an input credit at specific processing steps during remanufacturing. Progressive taxation can be difficult if the tax incidence is difficult to locate and if the cost of implementation is administratively high (Musgrave & Musgrave, 1973; Fernández-Albertos & Kuo, 2018). However, to combat this effect, we propose that this allowance be given for a fixed period of time and to a small number of companies who have the capability to setup and operate remanufacturing facilities, bigger than a certain size. Instead of offering this as an umbrella tax cut, marking down the audience of the tax allowance, will help discourage rent seeking activities as well as ensure that the remanufacturers who are meant to get this benefit do not falter, as their reputation is at stake. We label these as 'strategic tax allowances' as they are not quite given to a broad audience and are fairly specific in their purpose and their incidence.

6.3.1.2   Limited Time Discounts for Facilities

Hammond *et al.* (1998) surveyed several remanufacturers in Automotive Parts Rebuilders Associations Electrical Clinic in Orlando, USA about the difficulties in remanufacturing and setting up of a facility for remanufacturing was the most cumbersome and expensive process according to them. For this issue, we propose that at initial period exponentially decaying discount should be given to remanufacturing facilities for their land procurement. A rate can be fixed in time period $T = 0$ and we can calculate a downward cascading discount, which has an exponential decay with each passing year. So, the manufacturers who open their remanufacturing facility in $T = 3$, for instance, get a substantially lower discount on the land as the ones who setup the facility sooner. This will serve two important purposes – *firstly,* it implements the concept of 'time value of land', where land is treated as a finite and exigent resource and is therefore not disproportionately exploited. *Secondly,* it ensures that there is greater incentive to *start remanufacturing sooner than later.*

6.3.2   THE STATE–CONSUMER LINK

Spurling *et al.* (2013) have identified several ways to frame a problem to encourage pro-sustainability behaviour. Their approach theorizes that individual's behaviours are performances of social practices. According to the authors, social practices are made up of 'materials, components and meanings' (Spurling *et al.*, 2013), and it is these practices that need to be carefully observed and targeted through a nudge. They subscribe to a two-pronged approach for this: changing sequences of practices and synchronizing practices. For instance, introducing a remanufactured car in the market would require buyers to change the sequences of their purchasing practices and decision making and would also require several buyers to together accept this as a 'synchronized practice'. It will need to acquire enough cognitive legitimacy to become a default option while purchasing cars. However, we propose a modification in this approach of observing only individual behaviours and practices to refocus it onto community behaviours and practices. Creating an environment where a remanufactured car is considered as legitimate as a new car requires policy makers to target social norms, which purchases such as a car indicate in a society. For instance, the possession of a car is typically indicative of an increase in individual or family income. Similarly, in the Indian context, it is indicative of modernity and a rise in the income class of an individual, which signals a rise in their social status. Thus, for a policy maker to address embedded social practices, we propose that the nudge be at different levels of community engagement. In our recommendation, we posit that the nudges be given at three levels: the individual, the meso and the macro levels.

At the individual level, we propose to have nudges that create positive self-images about pro-social and sustainable behaviour. These can be through advertisements, campaigns to engage with volunteers and individuals who are passionate about the cause. This can also entail engaging universities and expert institutes doing technical research on the subject. At the meso level nudging would entail collaborating with civil society organizations such as social enterprises and NGOs. Numerous social enterprises exist today for transport optimization.

In stark contrast to the ownership model, the mobility as service (MaaS) model has transformed the way people travel. MaaS leverages digital technologies to enable end-to-end trip planning, booking and payment services across transport modes (Giesecke *et al.*, 2016) It includes ride hailing aggregators like Uber and Ola and bike sharing services like Yulu and MyByk in India, which is growing at a CAGR of 9.7% till 2025 with a total projected shared mobility fleet of 4.7 million units (Frost & Sullivan, 2019).

Although based in other geographies, there is evidence that prevalence of use of ride hailing services increases the overall vehicle miles travelled by a significant margin (Henao & Marshall, 2019). The MaaS market can serve as an attractive market for remanufactured vehicles, keeping in view deals and partnerships undertaken by some players with auto manufacturers (Thakkar, 2016; PTI, 2018).

Encouraging aggregators to favour remanufactured vehicles to a certain proportion of their fleet will lead to a change in the perceived quality and image of these vehicles. At the macro level, a large-scale nudge approach may be designed for all vehicle intensive consumer companies such as private road transport, agriculture, cargo-carriers, packers and movers, etc. These can be pushed through agencies such as the Confederation of Indian Industries (CII), Automotive Research Association of India, Central Public Works Department, etc. All government organizations and private firms, which employ automobiles at a large scale can be targeted through the nudges at this level.

A more direct approach would be through designing choice architecture in the form of simple interventions such as promoting remanufactured vehicles at all Passport offices and Regional Transport Authorities/Offices (RTAs/RTOs), where people get their driving licenses approved. Placing remanufactured cars in the same facilities as brand new cars would also help consumers make choices that might be equally effective for transportation needs and more environmentally conscious with similar brand values. OEMs might find this difficult due to fear of cannibalization, which we discuss in our final section on 'firm–remanufacturer' relationship. Another aspect of this direct approach can be a 'remanufactured vehicles only' policy for all government offices and for usage in all central and state government activities including for office use, field use and for usage by all levels of bureaucrats. Such initiatives could help promote a circular use pattern for consumers hesitant to try out remanufactured products.

These nudges must be accompanied by information on making the choice to buy a remanufactured car and its implications for the consumer. However, it is also important that the nudges be subtle and not come across as incentives. One reason that purely incentive-based approaches are not the option of choice is because they often cause distortions in consumer preferences, thus giving rise to inflation within the industry. Such distortions have far reaching consequences both in fiscal and monetary terms. Benartzi *et al.* (2017) compare nudge based and traditional tools in varied domains like public health, energy and education, and find nudges to be more impactful than incentive-based schemes. Thus, hybrid approach of nudges with tax-based incentives is more subtle and less invasive than only an incentive and provide a calculated impetus toward a particular choice (Schubert, 2017)

6.3.3 THE FIRM–CONSUMER LINK

The sub-discipline of environmental marketing holds much promise in 'nudging' consumers towards re-consumption and reforming their consumption behaviour by eliminating the negative connotations attached with a 'used' product. Marketers need to understand the cognitive biases and heuristics (Frederiks *et al.*, 2015) that affect the purchase process of a 'green' product and target the psychological determinants using advertising and promotion. Intensive market research will be useful in converting intention to action when it comes to buying green products. Each element of the marketing mix: product, packaging, positioning and promotion should be aligned to the goal reducing ecological damage (Polonsky, 2011). Panda *et al.* (2020) show a significant positive relationship between sustainability awareness and altruistic values, which positively drive consumer attitude and purchase intention for 'green' products, indicating that there is a customer segment with inclination towards responsible consumption (Prakash *et al.*, 2019).

One type of the outcome of a hybrid system are refurbished products that are often available at a lower cost than the newly manufactured products due to lower cost of production and material (Chen & Chang, 2013). This helps the firm expand its base to functionality oriented consumers who were earlier not being captured due to high price. Thus, adoption of a hybrid system helps in conversion of consumer need to a purchase, thus widening the consumer base and avoiding cannibalization of new products (Atasu *et al.*, 2010). Consumers, by expressing demand, would act as a stimulus for firms to continue their remanufacturing efforts and maintain product lines associated with it.

This creates a bi-directional flow of incentive in the consumer–firm relationship. While remanufactured products cater to environment-conscious consumers, the products must perform competitively on non-environmental attributes as well as shown by research in emerging markets context (Sonnenberg *et al.*, 2014).

6.4 THE AUTOMOBILE FIRM'S PERSPECTIVE: CHALLENGES AND OPPORTUNITIES

This section gives an extensive overview of literature on the choices faced by the automobile firm in setting remanufacturing operations.

6.4.1 Challenges in Operations Management

A firm planning to transition to a hybrid closed-loop system faces challenges with respect to inventory management, yield management, quality assessment and competition from other firms. These challenges are important in maintaining the slow adoption of remanufacturing by firms as they have conventionally operation with a 'linear' model, hence setting a reverse logistics chain is both complex and costly in nature. The operations management literature is replete with works that study hybrid systems using analytical and simulation techniques and offer solutions in terms of decision frameworks and guidelines. The inapplicability to repair systems literature due the assumption of one-one correspondence between returns and demand is in contrast to hybrid systems where the decisions between push/pull and hybrid/simple systems depend majorly on coordination level between demands and return. Push strategy means the returned products are processed as soon as they arrive, so the production schedule is returns driven. Pull strategy means the returned products are put as farther in the schedule as possible, which means the production schedule depends on returns, demand and serviceable inventory position. The lack of information about returns pattern and quality introduces variability in the system (Guide & Van Wassenhove, 2009). Supply chain process variability has been found to be detrimental to financial performance (Germain et al., 2008). Firms should invest in data collection capabilities by placing 'checkpoints' in the forward supply chain that helps in returns distribution by considering average lifetime and usage patterns. Big data capabilities has been found to support value creation in supply chain through expansive use of data, accelerating innovation in the long term (Wang et al., 2016; Lee, 2018).

Panagiotidou et al. (2017) derive value of information under different returns quality patterns and found it to be significant. This serves as a substantive economic rationale for firms that have hybrid systems but haven't invested in a quality assessment system for returned products. A dynamic quality assessment system can also help reduce system variance that stems from uncertain product quality. It also highlights a value-adding use of RFID data captured by many organizations. While setting up a reverse logistics chain, the firm has an option of either collecting returned products directly from consumers or route them through retailers. The latter has been shown to be an economical alternative (Hong et al., 2015). Table 6.1 summarizes the operational choices faced by firms while setting remanufacturing operations and associated literature. Adopting remanufacturing involves more than gearing the supply chain to handle returns. The firm needs to recognize people-related issues like incentive alignment that helps optimize the 'new objective function' (Kleindorfer et al., 2005). Since it is complicated to evaluate the capital tied in a product (Tang et al., 2004), bullwhip index can be used as a measure of hybrid system performance as system variance is one of the major concerns while adopting a hybrid approach (Dominguez et al., 2019). Operations management literature has borrowed extensively from information theory to compare different supply chain structures and strategies, which can be leveraged for evaluating reverse supply chains (Gravier & Kelly, 2012).

TABLE 6.1
Considerations in Setting Remanufacturing Operations

Considerations	Related Literature
Return pattern	Invest in data collection capabilities, statistical estimation (DeCroix & Zipkin, 2005)
Inventory policy	Pull reduces system variance (DeCroix & Zipkin, 2005)
Imperfect quality	Invest in quality assessment system (Panagiotidou *et al.*, 2017)
Minimize inventory cost	Economic order quantity (Teunter, 2001)
Reverse logistics chain operation	Contract collection activity to retailer (Hong *et al.*, 2015)

Source: Self sourced summary of literature on operational considerations for firms considering getting involved with remanufacturing of automotive and auto-parts.

6.4.2 THE FIRM–REMANUFACTURER RELATIONSHIP

While designing efficient production facilities and inventory management systems is a problem of local interest to a firm, it tends to overlook the exogenous factors like legislations and competition that might affect decisions like whether to remanufacture, how much to remanufacture and how to effectively manage the reverse logistics chain. This is an important consideration as Xerox and Deere & Co. have faced challenges in this context but have managed to set up successful remanufacturing programs (Majumder & Groenevelt, 2001). Setting up a new remanufacturing facility or tuning existing processes to handle returned products requires significant effort that is done post a thorough economic analysis. But presence of local remanufacturers severely impacts the quantity of products returned to the original manufacturer's facility, thereby disturbing the economic logic of remanufacturing. Webster & Mitra (2007) show that although presence of local remanufacturing units spurs remanufacturing activity but it is advisable for the OEM to purchase returns from the local units to maintain its competitive edge. While a number of options exist to cater to the supply chain, we recommend some state sponsored initiatives or public–private partnerships in certain areas that can help remanufacturing facilities and supply chains to become strong, locally rooted and independent.

In order to consolidate the supply chain and bring on board a skilled workforce as well as better design and quality control, we suggest, herewith, some measures that can potentially be incorporated into industry practices, through governance-practice linkages.

6.4.2.1 Setting up Remanufacturing Facilities Through Infrastructural Development

An Associated Chambers of Commerce and Industry of India (ASSOCHAM) remanufacturing seminar in 2009 concluded the key observation that OEMs should be allowed and incentivized to build remanufacturing facilities first, as they have the largest market shares in the Indian automobile market (Vasudevan *et al.* 2014). However, for fear of cannibalization, OEMs may not be keen on establishing their remanufacturing supply chains. In another study, Vasudevan *et al.* (2014) observe that out of the different reverse logistics mechanisms followed by companies such as HP, Xerox and Kodak in India, auto parts and automobile reverse logistics mostly happen through a third party, which consolidates parts from consumers and/or retailers and supplies them to remanufacturers (which maybe OEMs in some cases such as Volvo). We thus propose a twofold intervention through state agencies. For encouraging OEMs to invest in remanufacturing facilities, state governments may consider allocating lad for infrastructure development, that is close to OEM facilities in Special Economic Zones (SEZ) – such as the ELCOT SEZ in Chennai, which houses Mahindra's 'Auto Ancillary SEZ. Several exporters send their parts to this SEZ, where Mahindra houses skilled staff and remanufacturing infrastructure for auto parts manufacturing and assembly. We propose a similar SEZ-based model, where import duties on certain necessary remanufacturing parts can be reduced in order to have a smooth reassembly process.

6.4.2.2   Involving and Integrating Unorganized Efforts in the Supply Chain

A number of third parties involved in the remanufacturing process are unorganized and scattered in informal setups, from which, sourcing and transportation costs have to be borne by the OEMs. Due to this reason, a number of spare parts also end up in landfills (Vasudevan *et al.*, 2014). For this, we propose a productivity-based subsidy on transportation of recovered automotive parts/automobiles. We suggest that in exchange for housing energy efficient and quality-controlled remanufacturing facilities, the OEM company may claim rebate of road transportation taxes, conditional to their reported energy savings crossing a stringent minimum threshold in terms operation and productivity (this maybe defined through production efficiency or quality, durability and emissions of finished product). Thus by tying together transportation with the energy efficiency of remanufacturing processes, we hope to establish an 'emissions trade-off', where a fixed level of emissions is allowed within the entire collection–reassembly–remanufacturing process, and a respite is offered in one part of the cycle, in exchange for a calculated energy savings in another part, inspired by carbon credit systems.

6.4.2.3   Promoting Local Semi-Skilled Labour

One of the main barriers for successful remanufacturing effort on a large scale, which is identified in several studies is the lack of skilled labour (Vasudevan *et al.*, 2014; Govindan *et al.*, 2016; Chakraborty *et al.*, 2019). The (re)skilling of labour, drivers of technological growth, is a primary project on which the National Skill Development Mission (NSDM) in the country bases itself. The NSDM offers public private partnerships (PPPs) to place and train semi-skilled and skilled labour from all geographic backgrounds (rural, semi-urban and urban). These PPPs are potential source of skilled labour as well as equal employment opportunities for the remanufacturing sector. We propose the exploration of the NSDM as a road to developing long-term employment opportunities and income security for the work-force that trains under the OEM's facilities.

6.5 CONCLUSION

The concept of circular economy has gained much traction among academics and practitioners alike as its implementation holds the promise of a cleaner and greener environment. But the practice of studying the various stakeholders in silos has impeded the implementation of sectoral policies in a synergistic manner. Though this a consolidated framework, with the state agencies playing an important part in the processes within the supply chain and the marketing and distribution of remanu-factured vehicles, we expect to see a shift in both consumer mentality as well as industrial practices.

Common barriers to remanufacturing in emerging economies in various sectors, including the Indian automobile sector, have been substantively explored through various academic studies (Rashid *et al.*, 2013; Govindan *et al.*, 2016). However, a governance oriented practice framework, delineates transparent workable strategies for state intervention as well as provides a thoroughfare pathway to standardiza-tion. Currently, the auto parts and automobile remanufacturing industry in India is in a nascent stage and standard procedures for each part of the supply chain, from collection to reassembly to quality control are rudimentary as compared with other sectors such as photocopiers and camera remanufacturing, where initiative by indi-vidual industry leaders has resulted in sustainable outcomes (Mukherjee & Mondal, 2009; Chakraborty *et al.*, 2019). The framework suggested in this chapter is an effort to orient a cooperation-based strategy for a circular economy for the automobile industry. What makes this industry interesting is not only that it is perceived to be a large revenue generator for the Indian economy but also that it has been facing a steady recession and is need of innovative ideas for revival (Automotive Mission Plan, 2016–2026).

In this scenario, it is imperative that reworked imaginations of the standard solu-tions (tax rebates, special economic zones, reduction of import duties or easy land clearances) be explored for the state as well as industry to have steady and sus-tainable growth in the foreseeable future. Our idea of a modified 'carbon-credit' or 'energy-credit' system as rebate on the transportation part of the supply chain serves to connect issues of supply chain, and quality control, through a novel system of 'energy efficiency credits'. Similarly, the nudge-based approach for promotion of remanufactured vehicles will help drive a change in consumer mindset, while a higher investment in Big Data analytics will help develop resource planning and supply chain systems, to help maintain inventory and keep track of every step of the remanufacturing process. Thus what we propose here is a systemic approach to the involvement of policy and industry, through a collective endeavour. It requires a reimagining of fixated roles of the government and the firm but such cooperation based models are likely to bring in diverse strengths of individual actors and offers a holistic proposition for the automobile industry. The framework can also be applied to others sectors such as fast fashion and consumer electronics where waste genera-tion is a major concern for sustainable operations.

ENDNOTE

1 This excerpt is directly taken from the Department of Heavy industry website on their "Automotice Industry" webpage. Retrieved from the following URL: https://dhi.nic.in/ UserView/index?mid=1319

REFERENCES

Atasu, A., Guide, V. D. R., & Van Wassenhove, L. N. (2010). So what if remanufacturing cannibalizes my new product sales? *California Management Review, 52*(2), 56–76.

Amelia, L., Wahab, D. A., Haron, C. C., Muhamad, N., & Azhari, C. H. (2009). Initiating automotive component reuse in Malaysia. *Journal of Cleaner Production, 17*(17), 1572–1579.

Automotive Mission Plan, 2016–2026. Department of Heavy Industries, Government of India, January, 2016. Retrieved from: https://dhi.nic.in/writereaddata/Content/AMP%20 2016-26%20Final%20Approved%20Draft.pdf

Barker, S., & King, A. (2007). Organizing reuse managing the process of design for remanufacture (DFR). *18th Production and Operations Management Society Conference (POMS).* May 4th to May 7th, 2007. Retrieved from: https://www.poms.org/ conferences/poms2007/CDProgram/Topics/full_length_papers_files/007- 0769.pdf

Benartzi, S., Beshears, J., Milkman, K. L., Sunstein, C. R., Thaler, R. H., Shankar, M., … &Galing, S. (2017). Should governments invest more in nudging?. *Psychological Science, 28*(8), 1041–1055.

Bras, B. (2014). Design for remanufacturing processes. *Mechanical engineers' handbook*, (pp. 1–28). Hoboken, New Jersey: John Wiley & Sons.

DeCroix, G. A., & Zipkin, P. H. (2005). Inventory management for an assembly system with product or component returns. *Management Science, 51*(8), 1250–1265.

Dominguez, R., Cannella, S., Ponte, B., & Framinan, J. M. (2020). On the dynamics of closed-loop supply chains under remanufacturing lead time variability. *Omega, 97,* 102–106.

Chakraborty, K., Mondal, S., & Mukherjee, K. (2019). Critical analysis of enablers and barriers in extension of useful life of automotive products through remanufacturing. *Journal of Cleaner Production, 227,* 1117–1135.

Chen, J. M., & Chang, C. I. (2013). Dynamic pricing for new and remanufactured products in a closed-loop supply chain. *International Journal of Production Economics, 146*(1), 153–160.

Ellen MacArthur Foundation (2016). Circular Economy in India: Rethinking growth for long-term prosperity. Retrieved from: http://www.ellenmacarthurfoundation.org/ publications/

Fernández-Albertos, J., & Kuo, A. (2018). Income perception, information, and progressive taxation: Evidence from a survey experiment. *Political Science Research and Methods, 6*(1), 83–110.

Frederiks, E. R., Stenner, K., & Hobman, E. V. (2015). Household energy use: Applying behavioural economics to understand consumer decision-making and behaviour. *Renewable and Sustainable Energy Reviews, 41,* 1385–1394.

Frost & Sullivan (2019). With Projected CAGR of 9.7% over 2019–2025, Shared Mobility Set to Emerge as Major Transportation Mode across India. Retrieved from https:// ww2.frost.com/frost-perspectives/with-projected-cagr-of-9-7-over-2019-2025-shared-mobility-set-to-emerge-as-major-transportation-mode-across-india/

Germain, R., Claycomb, C., & Dröge, C. (2008). Supply chain variability, organizational structure, and performance: The moderating effect of demand unpredictability. *Journal of Operations Management, 26*(5), 557–570.

Giesecke, R., Surakka, T., & Hakonen, M. (2016, April). Conceptualising mobility as a service. In *2016 Eleventh International Conference on Ecological Vehicles and Renewable Energies (EVER)* (pp. 1–11). IEEE.

Govindan, K., Shankar, K. M., & Kannan, D. (2016). Application of fuzzy analytic network process for barrier evaluation in automotive parts remanufacturing towards cleaner production – A study in an Indian scenario. *Journal of Cleaner Production, 114,* 199–213.

Gravier M.J., Kelly B.P. (2012) Measuring the Cost of Complexity in Supply Chains: Comparison of Weighted Entropy and the Bullwhip Effect Index. In Jodlbauer H., Olhager J., Schonberger R. (eds) *Modelling Value. Contributions to Management Science.* Physica-Verlag HD. https://doi.org/10.1007/978-3-7908-2747-7_13

Guide Jr, V. D. R., & Van Wassenhove, L. N. (2009). OR FORUM—The evolution of closed-loop supply chain research. *Operations Research, 57*(1), 10–18.

Hammond, R., Amezquita, T., & Bras, B. (1998). Issues in the automotive parts remanufacturing industry: A discussion of results from surveys performed among remanufacturers. *Engineering Design and Automation, 4,* 27–46.

Henao, A., & Marshall, W. E. (2019). The impact of ride-hailing on vehicle miles travelled. *Transportation, 46*(6), 2173–2194.

Hong, X., Xu, L., Du, P., & Wang, W. (2015). Joint advertising, pricing and collection decisions in a closed-loop supply chain. *International Journal of Production Economics, 167,* 12–22.

Khan S.A.R., Jian C., Yu Z., Golpîra H., Kumar A. (2019) Impact of Green Practices on Pakistani Manufacturing Firm Performance: A Path Analysis Using Structural Equation Modeling. In: Anandakumar H., Arulmurugan R., Onn C. (eds) *Computational Intelligence and Sustainable Systems. EAI/Springer Innovations in Communication and Computing.* Cham, Springer. https://doi.org/10.1007/978-3-030-02674-5_6

Khan, S. A. R., Sharif, A., Golpîra, H., & Kumar, A. (2019b). A green ideology in Asian emerging economies: From environmental policy and sustainable development. *Sustainable Development, 27*(6), 1063–1075.

Kim, H. J., Raichur, V., & Skerlos, S. J. (2008, January). Economic and environmental assessment of automotive remanufacturing: Alternator case study. In *ASME 2008 International Manufacturing Science and Engineering Conference collocated with the 3rd JSME/ASME International Conference on Materials and Processing* (pp. 33–40). American Society of Mechanical Engineers Digital Collection. Retrieved from: http://css.umich.edu/sites/default/files/publication/CSS08-27.pdf

Kleindorfer, P. R., Singhal, K., & Wassenhove, L. N. (2005). Sustainable operations management. *Production and Operations Management, 14*(4), 482–492.

Lee, H. L. (2018). Big data and the innovation cycle. *Production and Operations Management, 27*(9), 1642–1646.

Lehner, M., Mont, O., & Heiskanen, E. (2016). Nudging – A promising tool for sustainable consumption behaviour? *Journal of Cleaner Production, 134,* 166–177.

Majumder, P., & Groenevelt, H. (2001). Competition in remanufacturing. *Production and Operations Management, 10*(2), 125–141.

Mangla, S. K., Luthra, S., Jakhar, S., Gandhi, S., Muduli, K., & Kumar, A. (2020). A step to clean energy – Sustainability in energy system management in an emerging economy context. *Journal of Cleaner Production, 242,* 118462.

Make in India (2015). *Sector Survey-Automobiles: India on its way to become the primary global automobile manufacturer.* Retrieved from https://www.makeinindia.com/article/-/v/make-in-india-sector-survey-automobile

McKenzie-Mohr, D. (2011). *Fostering sustainable behavior: An introduction to community-based social marketing.* Gabriola, Canada: New Society Publishers.

Mezzadri, A. (2014). Indian garment clusters and CSR norms: Incompatible agendas at the bottom of the garment commodity chain. *Oxford Development Studies*, *42*(2), 238–258.

Mukherjee, K., & Mondal, S. (2009). Analysis of issues relating to remanufacturing technology – A case of an Indian company. *Technology Analysis & Strategic Management*, *21*(5), 639–652.

Musgrave, R. A., & Musgrave, P. B. (1973). *Public finance in theory and practice*. New York City: NY: McGraw Hill Inc.

Panagiotidou, S., Nenes, G., Zikopoulos, C., & Tagaras, G. (2017). Joint optimization of manufacturing/remanufacturing lot sizes under imperfect information on returns quality. *European Journal of Operational Research*, *258*(2), 537–551. http://doi.org/10.1016/j.ejor.2016.08.044

Panda, T. K., Kumar, A., Jakhar, S., Luthra, S., Garza-Reyes, J. A., Kazancoglu, I., & Nayak, S. S. (2020). Social and environmental sustainability model on consumers' altruism, green purchase intention, green brand loyalty and evangelism. *Journal of Cleaner Production*, *243*, 118575.

Polonsky, M. J. (2011). Transformative green marketing: Impediments and opportunities. *Journal of Business Research*, *64*(12), 1311–1319.

Prakash, G., Choudhary, S., Kumar, A., Garza-Reyes, J. A., Khan, S. A. R., & Panda, T. K. (2019). Do altruistic and egoistic values influence consumers' attitudes and purchase intentions towards eco-friendly packaged products? An empirical investigation. *Journal of Retailing and Consumer Services*, *50*, 163–169.

Press Trust of India (2018). Honda signs up Drivezy for Its 2 Wheelers in Hyd, B'luru. *The Economic Times*, August 31st, 2018. Retrieved from https://auto.economictimes.indiatimes.com/news/two-wheelers/motorcycles/honda-signs-up-drivezy-for-its-2-wheelers-in-hyd-bluru/65626516

Rashid, A., Asif, F. M., Krajnik, P., & Nicolescu, C. M. (2013). Resource conservative manufacturing: An essential change in business and technology paradigm for sustainable manufacturing. *Journal of Cleaner production*, *57*, 166–177.

Sakai, S. I., Yoshida, H., Hiratsuka, J., Vandecasteele, C., Kohlmeyer, R., Rotter, V. S., ... & Oh, G. J. (2014). An international comparative study of end-of-life vehicle (ELV) recycling systems. *Journal of Material Cycles and Waste Management*, *16*(1), 1–20.

Schubert, C. (2017). Green nudges: Do they work? Are they ethical? *Ecological Economics*, *132*, 329–342.

Sonnenberg, N. C., Erasmus, A. C., & Schreuder, A. (2014). Consumers' preferences for eco-friendly appliances in an emerging market context. *International Journal of Consumer Studies*, *38*(5), 559–569.

Spurling, N.J., McMeekin, A., Southerton, D., Shove, E A., Welch, D. (2013) *Interventions in practice:reframing policy approaches to consumer behaviour*. Sustainable Practices Research Group, Manchester. https://eprints.lancs.ac.uk/id/eprint/85608/1/sprg_report_sept_2013.pdf

Subramoniam, R., Huisingh, D., & Chinnam, R. B. (2009). Remanufacturing for the automotive aftermarket-strategic factors: Literature review and future research needs. *Journal of Cleaner Production*, *17*(13), 1163–1174.

Sunstein, C. R. (2014). Nudging: A very short guide. *Journal of Consumer Policy*, *37*(4), 583–588.

Sutherland, J. W., Adler, D. P., Haapala, K. R., & Kumar, V. (2008). A comparison of manufacturing and remanufacturing energy intensities with application to diesel engine production. *CIRP Annals*, *57*(1), 5–8.

Tang, O., Grubbström, R. W., & Zanoni, S. (2004). Economic evaluation of disassembly processes in remanufacturing systems. *International Journal of Production Research*, *42*(17), 3603–3617.

Teunter, R. H. (2001). Economic ordering quantities for recoverable item inventory systems. *Naval Research Logistics (NRL)*, *48*(6), 484–495.

Thakkar, K. (2016, September 9). Mahindra & Mahindra inks 40,000 car supply deal with Ola. *The Economic Times*, Retrieved from https://economictimes.indiatimes.com/mahindra-mahindra-inks-40000-car-supply-deal-with-ola/articleshow/54169015.cms?from=mdr

Thaler, R., & Sunstein, C. (2008). *Nudge: The gentle power of choice architecture.* New Haven, Conn.: Yale.

Turaga, R. M. R., Howarth, R. B., & Borsuk, M. E. (2010). Pro-environmental behavior: Rational choice meets moral motivation. *Annals of the New York Academy of Sciences*, *1185*(1), 211–224.

Van Der Laan, E., Salomon, M., Dekker, R., & Van Wassenhove, L. (1999). Inventory control in hybrid systems with remanufacturing. *Management Science*, *45*(5), 733–747.

Vasanthagopal, R. (2011). GST in India: A big leap in the indirect taxation system. *International Journal of Trade, Economics and Finance*, *2*(2), 144.

Vasudevan, H., Kalamkar, V., Sunnapwar, V., & Terkar, R. (2014). Exploring the Potential of Remanufacturing in Indian Industries for Sustainability and Economic Growth. International Journal of Business Management & Research (IJBMR), 4(1), 25–38.

Wang, G., Gunasekaran, A., Ngai, E. W., & Papadopoulos, T. (2016). Big data analytics in logistics and supply chain management: Certain investigations for research and applications. *International Journal of Production Economics*, *176*, 98–110.

Webster, S., & Mitra, S. (2007). Competitive strategy in remanufacturing and the impact of take-back laws. *Journal of Operations Management*, *25*(6), 1123–1140.

Xiang, W., & Ming, C. (2011). Implementing extended producer responsibility: Vehicle remanufacturing in China. *Journal of Cleaner Production*, *19*(6–7), 680–686.

Zhang, J. H., Yang, B., & Chen, M. (2017). Challenges of the development for automotive parts remanufacturing in China. *Journal of Cleaner Production*, *140*, 1087–1094.

Section 3

Applications of Advanced Methods in the Adoption of Circular Economy Practices

7 Fuzzy Global Criterion Method for Solving Multiobjective Sustainable Supplier Selection Problem

Nurullah Umarusman
Faculty of Economics & Administrative Sciences,
Department of Business Administration,
Aksaray University, Aksaray, Turkey

CONTENTS

7.1   INTRODUCTION

The fast-developing technology and communication network helped to overcome the problems of communication between countries, and therefore the flow of information has turned into a global formation. Thus, with the cooperation between the businesses and the effective management of supply chain (SC), products and services can be delivered to the customer in a shorter time and demanding amounts (Acar & Çapkın, 2017). Businesses, which requires to ensure customer satisfaction by supplying products appropriate for customer needs in demand amounts, on the spot, on time and with minimum cost, have to form and manage the SC effectively. In this case, establishing SC management and its optimization gain importance (Güçlü & Özdemir, 2015). The SC term points to the interdependent network parties such as storage facilities, retailers, transporters, distributors and suppliers, which form a product's manufacturing and distribution channels from purchasing its raw materials to supplying it to the end user (Laínez-Aguirre & Puigjaner, 2015). Nonetheless, as environmental issues have risen in recent years, Conventional SCs have been considered responsible for producing large quantities of waste and emissions, and resource consumption leading to environmental degeneration and deteriorating environmental quality (Achillas et al., 2019). In recent conditions, conventional SC management has given place to Sustainable Supply Chain Management (S-SCM).

S-SCM is a capital, information, material and management that is related to economic, environmental, and social goals through cooperation with companies in the SC in terms of customer and shareholder needs (Seuring & Müller, 2008). Organizations had to attach importance to social and environmental factors besides economic factors, because with the pressures from stakeholders, it has become imperative to be included in the sustainability concept SC (Sağnak, 2020). Environmental, economic and social factors should be on businesses' agendas when preferring suppliers if sustainability is the matter. Integrating social and environmental criteria into economic aspects of company performance targets is a necessity along the entire SC (Bai & Sarkis, 2010). Although the environmental dimension is more prominent, it is essential to read the concept of sustainability as a whole of environmental, economic and social dimensions (Altuntaş & Türker, 2012).

A sustainable supplier is instrumental in creating a positive corporate image for buying businesses. Consequently, sustainability performance assessment of suppliers is vital for determining and selecting suitable suppliers (Khan et al., 2018). The selection of suitable suppliers have a crucial role in achieving the goals of the business, maintaining and strengthening the competitive level of the business in the long run. For this reason, supplier selection is a decision that is strategic for the business, and this decision should be made more scientifically and systematically, not based on intuition and experience (Çakın & Özdemir, 2013). The increasing attention for sustainability raises the challenges that decision makers face when selecting sustainable suppliers that regard environmental, economic and social aspects. In particular, decision makers are much more encouraged to improve their SC facilities to deal efficiently with sustainable development objectives (Mohammed et al., 2019). Sustainable supplier selection is the method of selecting an organization's suitable supply partners with the most advantageous monetary value, thus reducing the numerous social and environmental impacts of its operations (Moheb-Alizadeh & Handfield, 2019). Supplier selection and evaluation process are quite complex, requiring interrelationship between two or more SC entities, and the process is multiple objective (Sivrikaya et al., 2015). In general, the issue of supplier selection comes under the department of purchasing. The purchasing department's vital goal is to obtain the right product from the suitable source, at the right cost, in the right quantity, and with the right quality at the right time. This needs effective decisions about the evaluation and selection of suppliers (Ware et al., 2012).

Due to the Covid-19 pandemic, production processes have been affected negatively in particular, or production has been decreased. Businesses that want to be under-affected by the pandemic have reorganized work shifts and taken new measures for workers' health. Another negative effect of the global Covid-19 pandemic is the delay in delivery times of the products purchased. Therefore, it is an expected situation that many businesses will look for new ways in supplier selection to meet the demands. In this study, a supplier selection problem (SSP) of a firm in the construction sector has been handled based on on-time delivery performance level of products (%), green product quality (%), and performance level of work safety and worker health applications (%) criteria, and Fuzzy GCM algorithm has been used as the solution of this problem.

7.2 SUPPLIER SELECTION CRITERIA IN SUSTAINABILITY PROCESS

The process of selecting the right supplier is very complicated because profitability and customer satisfaction directly affect the process. The supplier selection process has become a difficult task for the business since controversial criteria will be used while selecting the right supplier. According to Umarusman (2019), all the criteria for sustainable economic selection directly or indirectly follow the criteria defined by Dickson as a very crucial average value. This means conventional supplier selection usually applies to economic criteria. Technological advancements, diminishing natural resources with a growing population, environmental pollution harm to the ecosystem, and socio-cultural changes have taken the criteria that relate to the definition of sustainability in the selection and evaluation of suppliers to a more substantial level in the last 20 years. The first study on conventional Supplier Selection Criteria (SSC) was carried out by Dickson (1966). Afterward, from different perspectives, Dempsey (1978), Roa and Kiser (1980), Bache et al. (1987), Weber et al. (1991), Cheraghi et al. (2004), and Thiruchelvam and Tookey (2011) expanded SSC. Özçelik and Avcı Öztürk (2014), Grover et al. (2016), Ghoushchi et al. (2018), and Li et al. (2019) classified within the frame of Triple Bottom Line (TBL) for sustainable SSC. Humphreys et al. (2003) proposed a frame for environmental criteria to be included in the selection process and defined the criteria and sub-criteria. Increased environmental awareness led to the emergence of the Green SC paradigm, and thus green criteria were included in SSP (Genovese et al., 2010). According to Umarusman and Hacıvelioğulları (2020), Green SSC are compatible with TBL criteria when SC processes are examined, so S-SCM includes Green SCM. There are scientific studies on Green SSC that take part in literature. For instance, Chiou et al. (2008) listed the global criteria and sub-criteria weights for green supplier selection. Govindan et al. (2015) classified the Green SSC used within the framework of Multiple Criteria Decision Making (MCDM) as a result of its literature research. Chen et al. (2016) provides the criteria that is most widely used in the literature for evaluating environmental and economic performance of green supplier selection. Haeri and Rezaei (2019) determined criteria, within the frame of TBL, frequently used in literature.

It is expected to use different criteria on a sectoral basis while evaluating and selecting suppliers in SC processes. On the other hand, the favorableness level will increase with supplier evaluation and managing accurately the process of determining the most appropriate method in supplier selection.

7.2.1 Literature Survey

Many different methods are used in solving Conventional/Sustainable/Green SSP. Such problems were classified by Ghodsypour and O'Brien (1998), De Boer *et al.* (2001), Humphreys *et al.* (2003), Ding *et al.* (2005), Genovese *et al.* (2010), Ware *et al.* (2012), Chai *et al.* (2013), Govindan *et al.* (2015), Ayhan and Kilic (2015), Trisna *et al.* (2016), Mukherjee (2017), Ozyoruk (2018), Banasik (2018), Umarusman (2019) and Chai and Ngai (2019) from different viewpoints. In this study, a literature review that includes multiple objective decision making (MODM) and Fuzzy MODM methods was done considering scientific studies conducted between 2010 and 2020. In the literature review, besides methods used in related studies, Table 7.1 shows that objective/goal functions are created in which criteria.

According to the literature review, when an evaluation was made in terms of supplier selection type, it was determined that in the studies conducted between 2010

TABLE 7.1
The Classifications in Supply Chain Processes

Author(s)	Methods	Objective /Goal Functions	Supplier Selection Type
Osman and Demirli (2010)	Bilinear Goal Programming and Modified Benders Decomposition Algorithm	Delivery time and distribution cost	Conventional SSP
Liao and Kao (2010)	Taguchi Loss function, AHP and Multi-choice Goal Programming	Product quality, price, delivery time, service satisfaction and warranty degree	Conventional SSP
Ku *et al.* (2010)	Fuzzy AHP and Fuzzy GP	Cost, quality, service and risk	Conventional SSP
Yücel and Güneri (2011)	Fuzzy MOLP	Net price, quality and delivery time	Conventional SSP
Elahi *et al.* (2011)	Fuzzy Compromise Programming	Cost, delivery lateness and defective rate of purchased product	Conventional SSP
Ozkok and Tiryaki (2011)	Fuzzy MOLP and Werners' Approach	Total purchased cost, service quality and item quality	Conventional SSP
Shaw *et al.* (2012)	Fuzzy-AHP and Fuzzy MOLP	Satın alma maliyeti, kalite, number of late delivered, and the total carbon foot print of the purchased item	Conventional SSP
Aouadni *et al.* (2013)	Imprecise Goal Programming and AHP	Price, capacity, and delivery	Conventional SSP
Nazari-Shirkouhi *et al.* (2013)	Two-phase Fuzzy MOLP	Total purchasing and ordering cost, net number of rejected items, net number of late delivered items	Conventional SSP
Liu and Papageorgiou (2013)	ε-constraint method, Lexicographic method	'Total cost', 'total flow time' and 'total lost sales'	Conventional SSP

(Continued)

TABLE 7.1 *(Continued)*
The Classifications in Supply Chain Processes

Author(s)	Methods	Objective /Goal Functions	Supplier Selection Type
Arikan (2013)	Fuzzy additive model, Augmented max–min model	'Cost', 'quality' 'on-time delivery'	Conventional SSP
Sheikhalishahi and Torabi (2014)	Fuzzy/soft Lexicographic Goal Programming	'The total initial cost' 'total risk of purchasing' 'total downtime cost' and the unreliability of the system	Conventional SSP
Choudhary and Shankar (2014)	Linear Goal Programming Variants	'Net rejected items', 'net cost' and 'net late delivered items'	Conventional SSP
Ashlaghi (2014)	Fuzzy ANP, Fuzzy DEMATEL, and Linear Physical Programming	'Cost', quality, service, environmental	Green SSP
Kazemi et al. (2014)	TOPSIS, Interactive-Fuzzy MOLP, Weighted additive model	'Purchasing cost', delay time, defect rate and transportation cost, total value of purchasing	Conventional SSP
Jadidi et al. (2015)	Multi-choice goal programming	Price, 'rejects and lead-time'	Conventional SSP
Azadnia et al. (2014)	Fuzzy AHP, a weighted sum method and an augmented ε-constraint method	'Total cost', 'total social score', 'total environmental score' and 'total economic qualitative score'	Sustainable SSP
Moghaddam (2015)	Monte Carlo simulation, non-preemptive goal programming, compromise programming and fuzzy goal programming	Total profit, defective parts late delivery economic risk	Conventional SSP
Ayhan and Kilic (2015)	Fuzzy AHP and Mixed Integer Linear Programming	Quality, price, after sales perf., delivery time performance	Conventional SSP
Hu and Yu (2016)	Voting Method and the Goal Programming.	Total value of purchase and total cost of purchase	Conventional SSP
Şenocak ve Güner Gören (2016)	Fuzzy DEMATEL, fuzzy GRA and fuzzy linear programming	Value of purchasing	Sustainable SSP
Pandey et al. (2017)	Fuzzy goal programming	Economic aspect, lean practices, sustainability, services	Sustainable SSP
Yousefi et al. (2017)	Global Criterion Method	Visibility, delayed or defective parts, supply chain cost	Conventional SSP
Umarusman and Hacıvelioğulları (2018)	Global Criterion Method	Cost, reject (%), and service (%)	Conventional SSP
Ekhtiari et al. (2018)	Nadir Compromise Programming	Reliability, flexibility, quality, on-time delivery, waste percentage, price	Conventional SSP

(Continued)

TABLE 7.1 *(Continued)*
The Classifications in Supply Chain Processes

Author(s)	Methods	Objective /Goal Functions	Supplier Selection Type
Mirzaee et al. (2018)	Preemptive fuzzy goal programming	Total cost, value of purchasing	Conventional SSP
Çalık (2018)	AHP, Fuzzy MOLP	CO_2 emission, energy consumption, waste production	Green SSP
Lo et al. (2018)	BWM (best worst method) TOPSIS and Fuzzy MOLP	Cost, delivery performance, product quality and total utility	Green SSP
Ho (2019)	Weighted multi-choice goal programming and MINMAX Multi-Choice Goal Programming	Cost, quality, technology innovation, delivery, production capacity	SSP
Umarusman (2019)	Global criterion Method, Compromise Programming, Minmax Goal programming and STEP method	Quality (%), guarantee and compensation (%), and Product Unit Price	Sustainable SSP
Moheb-Alizadeh and Handfield (2019)	DEA ve ε-constraint method and Benders decomposition algorithm	Total cost, emission, social responsibility	Sustainable SSP
Rabieh et al. (2019)	Fuzzy TOPSIS, ε-constraint method, Multi-Objective Programming and Weighted Sum Method.	Total cost, non-cost economical criteria, environmental score, social performance	Sustainable SSP
Umarusman and Hacıvelioğulları (2020)	De Novo Programming, Compromise Programming, MOLP	Technical capability, quality, service	Green SSP
Kaviani et al. (2020)	Intuitionistic Fuzzy AHP, Fuzzy MOLP, the weighted fuzzy model	Total financial cost, the geographical distance from the suppliers, total quality of purchased units, on-time delivery rate, the background of the partnership with the suppliers and the reputation and credibility, respectively	Conventional SSP

and 2020, *Conventional SSP* was used as 70%, *Sustainable SSP* was used as 18%, and *Green SSP* was used as 12%. These rates are shown in Figure 7.1.

New classifications can be made regarding the criteria and methods used in supplier selection processes using the information provided in Table 7.1. A new assessment has not been made because of the studies above, in which both methods and criteria are reviewed in the literature. However, the results can be drawn as

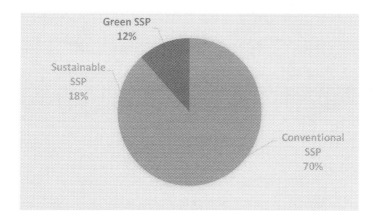

FIGURE 7.1   Type of supplier selection problems.

follows regarding the criteria and methods: Cost, Quality, Delivery Time criteria are much preferred in the establishment of Objective/Goal functions when Table 7.1 is analyzed.

7.2.2 FORMULATION OF MULTIPLE OBJECTIVE SUPPLIER SELECTION PROBLEM

Although the mathematical formulation of SSPs accepted as a special extension of Multiple Objective Linear Programming (MOLP) was realized by Gaballa (1974), Moore and Fearon (1973) discussed linear programming could be used in supplier selection for the first time. In literature, Weber and Ellram (1993), Ghodsypour and O'Brien (1998), Amid *et al.* (2011), and Umarusman (2019) organized the mathematical model of multiple objective SSP according to different objective functions. The arrangement made for three objective functions in the maximization type of multi objective supplier selection problem is given below in this study. The total budget has been used in restricting the amount to be purchased in this model.

$$\text{Max } Z_1 = \sum_{i=1}^{n} c_j^1 x_j$$

$$\text{Max } Z_2 = \sum_{i=1}^{n} c_j^2 x_j$$

$$\text{Max } Z_3 = \sum_{i=1}^{n} c_j^3 x_j$$

Subject to (1)

$$\sum_{i=1}^{n} f_i x_i \leq B$$

$$\sum_{i=1}^{n} x_i \geq D$$

$$x_i \leq C_i$$

$x \geq 0$ and integer.

Z_1, Z_2, Z_3: Maximization-typed objective function for the criteria,
B: Total budget,
f_i: Unit cost of the product to be purchased from ith supplier,
x_i: The amount to be purchased from i-th supplier,
D: Demand over period,
C_i: Capacity of ith supplier,
n: Number of the suppliers.

Equation (1) is accepted as MOLP problem, when investigating the solution, defining non-dominate solutions set is the first step. Moreover, the number of solutions in terms of non-dominate solutions set in (1) is quite a few. The basic process to evaluate non-dominate solution points is how the proximity to ideal solutions will be determined (Cohon, 1978, p. 69). Therefore, ideal solutions are designated at first. In (1), for each objective function, Positive Ideal Solution (PIS) set is $I^* = \left\{ Z_1^*, Z_2^*, \ldots Z_n^*; W_1^* W_2^*, \ldots, W_n^* \right\}$, and Negative Ideal Solution (NIS) is $I^- = \left\{ Z_1^-, Z_2^*, \ldots Z_n^*; W_1^-, W_2^-, \ldots, W_n^- \right\}$ (Zeleny, 1973). Afterward, using the methods in MODM classification, the solution of (1) is made. In the literature research given above, methods that can be used in the solution of (1), and integrated methods have been given. According to this literature research, it has been detected that the solution of Conventional/Sustainable/Green SSP is not made using the Fuzzy Global Criterion Method (Fuzzy GCM).

7.3   GLOBAL CRITERION METHOD

The first studies on the GCM, which is included in the classification of the 'preference information' of the MODM methods, were carried out by Yu (1973) and Zeleny (1973). GCM is given mathematically below

$$\text{Min} \sum_{i=1}^{n} \left(\frac{f_i\left(x^*\right) - f_i\left(x\right)}{f_i\left(x^*\right)} \right)^p$$

Subject to (2)

$$Ax \leq b$$

$$x \geq 0.$$

The benefit of this approach is that it is possible to derive the solution from the decision-making itself, and not from the designer (Shih & Chang, 1995). In this method, the distance is minimized between some reference points and the feasible objective region. Additionally, it is accepted that all objective functions have equal importance in the GCM (Miettinen, 2012). According to the method, more than one objective function is transformed into a single-objective optimization problem (Tabucanon, 1988). Within the global objective function, each objective function is expressed as a ratio. Non-dimensionalization is necessary because there are different dimensions to objective functions. The global objective value should be within the range of [0;1], as normalization is done on the basis of the constructive ideal solutions of objectives. The best solution for the problem varies according to the p-value chosen (Umarusman, 2019). Umarusman and Türkmen (2013) extended the GCM for problems with goals of maximization and minimization. From this point of view, GCM is given mathematically below.

$$\text{Min } G = \sum_{k=1}^{l}\left[\frac{Z_k^* - Z_k(x)}{Z_k^*}\right]^p + \sum_{s=1}^{r}\left[\frac{W_s(x) - W_s^*}{W_s^*}\right]^p$$

Subject to (2.1)

$$Ax \le b$$

$$x \ge 0.$$

Here;

$Z_k(x)$: Maximization-directed objective function,
Z_k^*: Positive ideal solution of k-th objective function,
$W_s(x)$: Minimization directed objective function,
W_s^*: Positive ideal solution of s-th objective function,
A: Matrix of technical coefficients,
b: Right-hand-side constants for the constraints,
p: $(1 \le p \le \infty)$.

Although a GCM can be a mathematical function with no association of preferences, a weighted GCM is a kind of utility function in which system parameters are used to model preferences (Marler & Arora, 2004). Weight factors for objective functions have not been used in this study.

7.3.1 Fuzzy Global Criterion Method

Fuzzy GCM could be seen as an extension of the GCM approach. Conflicting objectives resolve on the basis of the fuzzy-PIS only by deriving a collection of compromise solutions. In this case, a fuzzy set (Chang & Pires, 2015) is formed only by a PIS. A compromise solution is sought with respect to distance formulation by reducing the distance from a particular aspect of the fuzzy-PIS (Chang & Lu, 1997). This procedure has the benefit that the solution can be derived from the decision-making itself, not from the planner (Shih & Chang, 1995). The steps of the Fuzzy GCM proposed by Leung (1984) are as follows.

Step 1: Find the PIS $f^{1*} = (f_1^{1*}, f_2^{1*}, \dots, f_k^{1*})$ where f_k^{1*} is the solution of the problem below:

$$\max f_k(x)$$

Subject to (3)

$$g_i(x) \le b_i, \ \forall_i, \ x \ge 0.$$

where no tolerance value ü is considered.

Step 2: Acquire a compromise solution equivalent to f^{1*}. This compromise solution can be acquired by solving the problem below.

$$\min d_p^1 = \left\{ \sum_k \left[\frac{\left(f_k^{1*} - f_k(x)\right)}{f_k^{1*}} \right]^p \right\}^{1/p}$$

Subject to (4)

$$x \in X = \left\{ x \mid g_i(x) \le b_i, \ \forall_i, \ x \ge 0 \right\}$$

Where $\left[f_k^{1*} - f_k(x) \right] = \left[1 - \mu_k \left(f_k(x) \right) \right]$ if $\mu_k \left(f_k(x) \right) = \frac{f_k(x)}{f_k^{1*}}$ (here f_k^{1*} is crisp), and p can be any nature number.

For simplicity, we might only consider the following cases $p = 1$ and ∞.

$$\text{Min } d_1^1$$

Subject to (4.1)

$$x \in X$$

whose solution is assumed to be x^{11}, and

$$\text{Min } d_\infty^1$$

Subject to (4.2)

$$\frac{\left(f_k^{1*} - f_k(x) \right)}{f_k^{1*}} \le d_\infty^1$$

$$x \in X, \ \forall_k.$$

whose solution is assumed to be $x^{1\infty}$. Both compromise solutions x^{11} and $x^{1\infty}$ (f^{11} and $f^{1\infty}$) serve as bounds of the compromise solution set for $1 \le p < \infty$.

Step 3: Like Step 2, find the PIS within the constraints with maximum tolerances. That is, total avaible resources are $b_i + p_i \; \forall_k$. The PIS $f^{0*} = \left(f_1^{0*}, f_2^{0*}, \dots, f_k^{0*} \right)$ where f_k^{0*} is the solution of the problem below:

$$\text{Max } f_k(x)$$

Subject to (5)

$$g_i(x) \leq b_i + p_i, \; \forall_i, \; x \geq 0.$$

Step 4: Like Step 2, find the compromise solution with respect to f^{0*} solving the problem below:

$$\min d_p^0 = \left\{ \sum_k \left[\frac{\left(f_k^{0*} - f_k(x) \right)}{f_k^{0*}} \right]^p \right\}^{1/p}$$

Subject to (6)

$$x \in X = \left\{ x \mid g_i(x) \leq b_i + p_i, \; \forall_i, \; x \geq 0 \right\}$$

Again, we might just solve two boundary situations d_1 and d_∞ with the corresponding solution x^{01} and $x^{0\infty}$.

By stretching the limits of the constraints from b_i to $b_i + p_i, \forall_i$, maximum values of individual objectives functions $f_k(x), \forall_i$, shift from f_k^{1*} to f_k^{0*}. Consequently, the positive ideal solution moves from f_k^{1*} to f_k^{0*}. The positive idal becomes fuzzy and can be identidied as $\left\{ f_1(x), \dots, f_k(x), \left| f_k^{1*} \leq f_k(x) \leq f_k^{0*}, \; k = 1, \dots, K \right. \right\}$. For each single objective, we have $f_k(x) \geq \left[f_k^{1*}, f_k^{0*} \right], \; \forall_i$ within fuzzy constraints with various and different tolerance values.

Step 5: By solving the following K single-objective fuzzy linear programming obtain the most appropriate PIS.

$$f_k(x) \geq f_k^{1*}; f_k^{0*}$$

Subject to (7)

$$g_i(x) \leq b_i; b_i + p_i, \; \forall_i.,$$

For convenience, let's say membership functions are linear, with fuzzy objectives and constraints like:

$$\mu_k(x) = \begin{cases} 1 & \text{if } f_k^{0*} < f_k(x) \\ \left[1 - \left[f_k^{0*} - f_k(x) \right] \right] \Big/ \left(f_k^{0*} - f_k^{1*} \right) & \text{if } f_k^{1*} \leq f_k(x) \leq f_k^{0*} \\ 0 & \text{if } f_k(x) < f_k^{1*} \end{cases} \quad (8)$$

for $\forall_i$, and

$$\mu_i(x) = \left\{ 1 - \left[\begin{matrix} 1 \\ g_i(x) - b_i \\ 0 \end{matrix} \right] \right\} \Big/ p_i \quad \begin{matrix} if & g_i(x) < b_i \\ if & b_i \le g_i(x) \le b_i + p_i \\ if & g_i(x) > b_i + p_i \end{matrix}$$

for $\forall_i$. By using Bellman and Zadeh's max–min operator, we have (for each single objective):

$$\max_{x \in X} \min \left\{ \mu_k(f(x)), \mu_i(f(x)) \right\} \tag{9}$$

or

$$\max \alpha$$

Subject to (10)

$$1 - \left[f_k^{0*} - f_k(x) \right] \Big/ \left(f_k^{0*} - f_k^{1*} \right) \ge \alpha$$

$$1 - \left[g_i(x) - b_i \right] \Big/ p_i \ge \alpha$$

$$\alpha \in [0.1] \text{ and } x \ge 0$$

whose solution is assumed to be $f_k^* = f_k(x^*)$ and α_k^*. After solving K problems of Equation (10), the most suitable PIS can be $f^* = \left(f_1^*, f_2^*, \dots, f_k^* \right)$ with $\alpha^* = \max_k \alpha_k^*$.

Step 6: Acquire a compromise solution from original MOP when the most suitable PIS is f^* and α^*. With α^*, new constraints wil be

$$g_i(x) \le b_i + (1 - \alpha^*) p_i \tag{11}$$

Therefore, the compromise solutions of (1) can be obtained by using the global criterion (distance metric family) by solving the problem below:

$$\text{amax } d_p = \left\{ \sum_k \left[\frac{\left(f_k^* - f_k(x) \right)}{f_k^*} \right]^p \right\}^{1/p}$$

Subject to (13)

$$x \in X = \left\{ x \middle| g_i(x) \le b_i + (1 - \alpha^*) p_i, \forall_i, x \ge 0 \right\}$$

Where, p can be any nature number. Again, we may only chose to solve boundary situations of $p = 1$ and ∞. Leung (1984) stressed that such a fuzziness of constraints is versatile to allow one to create alternatives to resolve disputes that cannot otherwise be resolved or dissolved within permissible limits of tolerance. This explains that, in a dispute settlement, DMs don't need to confine themselves to a single point-valued PIS. The method of defining differing values and solutions to compromise is tractable.

7.4 APPLICATION

Akyamaç Engineering & Construction Firm in the construction sector in Aksaray, Turkey is planning on purchasing heat and noise insulation materials for a new project. Firm management is desiring to make a new buying plan based on three criteria from five different suppliers that it worked in the past period. The business top manager wants to be under-affected by delays during product delivery due to Covid-19. Therefore, the delivery time of the products in the last 8 months has been examined, and 'On-Time Delivery Performance Level of the Product (%)' has been determined as the first criteria for each supplier. In order to reduce the negative effects of the human and environmental health of the material to be purchased, the criteria of 'Green Product Quality' and 'Occupational Safety and Occupational Health Practices' applied by suppliers due to the epidemic have been scored as '%'. Information about criteria and unit price of the material to be purchased from suppliers are given in Table 7.2.

The business has detected that the demand for the products will be 450 at least and has budgeted $26,000 in total for the material to be purchased from the suppliers. Besides, according to previous period information of the business, as per agreements with suppliers, at least 100 from Supplier-1, at least 130 from Supplier-2, and at most 170 from Supplier-5 are purchased. A minimum of 140 and a maximum of 180 from Supplier-3, a minimum of 155, and a maximum of 165 from Supplier-4 can be purchased. The number of materials to be purchased from each supplier is given in Table 7.3.

Firm management is planning on increasing the amount to be purchased from Supplier-3, Supplier-4 and Supplier-5 according to the buying plan. The tolerance values for these suppliers are 15, 20, 30, respectively. The business has allocated an

TABLE 7.2
Performance Information of Suppliers

	Max Z_1: On-Time Delivery Performance Level of the Product (%)	Max Z_2: Green Product Quality (%)	Max Z_3: Work Safety and Labor Health (%)	Product Unit Cost (25 m²/$)
Supplier-1	75	80	90	35
Supplier-2	60	75	80	30
Supplier-3	80	70	70	38
Supplier-4	85	70	75	42
Supplier-5	70	80	100	40

TABLE 7.3
The Number of Materials to be Purchased from Suppliers

Suppliers	Minimum	Maximum
Supplier-1	100	-
Supplier-2	130	-
Supplier-3	140	180
Supplier-4	155	165
Supplier-5	-	170

additional budget, which is \$4000. According to the information given above, the Multiobjective Sustainable Supplier Selection problem (S-SSP) of the business is created as follows:

$$\text{Max } Z_1(x): 75x_1 + 60x_2 + 80x_3 + 70x_4 + 92x_5$$

$$\text{Max } Z_2(x): 89x_1 + 92x_2 + 95x_3 + 90x_4 + 0.95x_5$$

$$\text{Max } Z_3(x): 90x_1 + 90x_2 + 92x_3 + 85x_4 + 93x_5$$

Subject to (P1)

$$g_1(x): 35x_1 + 30x_2 + 38x_3 + 42x_4 + 40x_5 \leq 26,000$$

$$g_2(x): x_1 + x_2 + x_3 + x_4 + x_5 \geq 450$$

$$g_3(x): x_1 \geq 100; \ g_4(x): x_2 \geq 130$$

$$g_5(x): x_3 \geq 140; \ g_6(x): x_3 \leq 180$$

$$g_7(x): x_4 \geq 155; \ g_8(x): x_4 \leq 165$$

$$g_9(x): x_5 \geq 145; \ g_{10}(x): x_5 \leq 170$$

$$x_1, x_2, x_3, x_4, x_5 \geq 0 \text{ and integer.}$$

Step 1: In this step, the positive ideal solution set is determined. Solutions determined for each goal function in (P1) are given below.

$$\text{Max } Z_1(x); \ x_1 = 126; \ x_2 = 132; \ x_3 = 140; \ x_4 = 155; \ x_5 = 145; \ Z_1^{1*} = 51895$$

$$\text{Max } Z_2(x); \ x_1 = 102; \ x_2 = 160; \ x_3 = 140; \ x_4 = 155; \ x_5 = 145; \ Z_2^{1*} = 52410$$

$$\text{Max } Z_3(x); \ x_1 = 102; \ x_2 = 160; \ x_3 = 140; \ x_4 = 155; \ x_5 = 145; \ Z_3^{1*} = 57905$$

The ideal solution set of (P1) is obtained as $Z^{1*} = \{51,895;\ 52,410;\ 57,905\}$.
Step 2: In this step, compromised solutions are determined by depending on $Z^{1*} = \{51,895;\ 52,410;\ 57,905\}$. Namely, the solution is made for GCM (for $p = 1$) and Compromise Programming (CP) (for $p = \infty$). While determining the longest distance for $p = 1$, the longest distance is completely dominated for $p = \infty$.
  i. for $p = 1$; (P1) is organized as follows.

$$\text{Min } G:\ \big[3 - (0.004256x_1 + 0.003969x_2 + 0.004086x_3 + 0.004269x_4 + 0.004602x_5)\big]$$

In order global objective function to be minimum, $(0.004256x_1 + 0.003969x_2 + 0.004086x_3 + 0.004269x_4 + 0.004602x_5)$ should be maximum. According to this, the model below is obtained.

$$\text{Max } F:\ (0.004256x_1 + 0.003969x_2 + 0.004086x_3 + 0.004269x_4 + 0.004602x_5)$$

Subject to (P1.1a)

$$g_1(x):\ 35x_1 + 30x_2 + 38x_3 + 42x_4 + 40x_5 \le 26,000$$

$$g_2(x):\ x_1 + x_2 + x_3 + x_4 + x_5 \ge 450$$

$$g_3(x):\ x_1 \ge 100;\ g_4(x):\ x_2 \ge 130$$

$$g_5(x):\ x_3 \ge 140;\ g_6(x):\ x_3 \le 180$$

$$g_7(x):\ x_4 \ge 155;\ g_8(x):\ x_4 \le 165$$

$$g_9(x):\ x_5 \ge 145;\ g_{10}(x):\ x_5 \le 170$$

$$x_1, x_2, x_3, x_4, x_5 \ge 0 \text{ and integer.}$$

The variable values obtained from the solution of (P1.1a) are specified as $x_1 = 102$; $x_2 = 160$; $x_3\ 140$; $x_4 = 155$; $x_5 = 146$; Max $F = 2.99771$. Accordig to these variable values, $x^{11} = (102;\ 160;\ 140;\ 155;\ 146)$ ve $Z^{11} = (51,775;\ 52,410;\ 7,905)$ are obtained. Moreover, Min $G : \big[3 - (2.997717\,)\big]$ *is* 0.002283. Accordingly, the goal functions are very close to their positive ideal solutions.
  ii. for $p = \infty$; (P1) is organized as follows.

$$\text{Min } d_\infty$$

Subject to (P1.1b)

$$75x_1 + 60x_2 + 80x_3 + 70x_4 + 92x_5 + 51,895d_\infty \ge 51,895$$

$$89x_1 + 92x_2 + 95x_3 + 90x_4 + 0.95x_5 + 52,410d_\infty \geq 52,410$$

$$90x_1 + 90x_2 + 92x_3 + 85x_4 + 93x_5 + 57,905d_\infty \geq 57,905$$

$$g_1(x): 35x_1 + 30x_2 + 38x_3 + 42x_4 + 40x_5 \leq 26,000$$

$$g_2(x): x_1 + x_2 + x_3 + x_4 + x_5 \geq 450$$

$$g_3(x): x_1 \geq 100; \quad g_4(x): x_2 \geq 130$$

$$g_5(x): x_3 \geq 140; \quad g_6(x): x_3 \leq 180$$

$$g_7(x): x_4 \geq 155; \quad g_8(x): x_4 \leq 165$$

$$g_9(x): x_5 \geq 145; \quad g_{10}(x): x_5 \leq 170$$

$$x_1, x_2, x_3, x_4, x_5 \geq 0 \text{ and integer.}$$

The variable values and objective function values specified from the solution of (P1.1b) are obtained as $x^{1\infty} = (114; 146; 140; 155; 14.$ and $Z^{1\infty} = (51,835; 52,320; 57,865)$. Besides, Min $d_\infty = 0.001717$ shows that objective functions are close to their own positive ideal solutions.

Step 3: Similar to Step 1, find the PIS under the constraints with maximum tolerances. Maximum amounts to be purchased from Supplier-3, Supplier-4 and Supplier-5 are demanded to be increased as 15, 20 and 30, respectively. Due to these increases, the budget has been increased $4000. According to these information, (P1) is organized as follows.

$$\text{Max } Z_1(x): 75x_1 + 60x_2 + 80x_3 + 70x_4 + 92x_5$$

$$\text{Max } Z_2(x): 89x_1 + 92x_2 + 95x_3 + 90x_4 + 0.95x_5$$

$$\text{Max } Z_3(x): 90x_1 + 90x_2 + 92x_3 + 85x_4 + 93x_5$$

Subject to (P2)

$$g_1(x): 35x_1 + 30x_2 + 38x_3 + 42x_4 + 40x_5 \leq 30,000$$

$$g_2(x): x_1 + x_2 + x_3 + x_4 + x_5 \geq 450$$

$$g_3(x): x_1 \geq 100; \quad g_4(x): x_2 \geq 130$$

$$g_5(x): x_3 \geq 140; \quad g_6(x): x_3 \leq 195$$

$$g_7(x): x_4 \geq 155; \quad g_8(x): x_4 \leq 185$$

$$g_9(x): x_5 \geq 145; \ g_{10}(x): x_5 \leq 200$$

$$x_1, x_2, x_3, x_4, x_5 \geq 0 \text{ and integer.}$$

In (P2), solutions made for each objective function are given below.

Max $Z_1(x); \ x_1 = 242; \ x_2 = 130; \ x_3 = 140; \ x_4 = 155; \ x_5 = 145; \ Z_1^{0^*} = 60,475$

Max $Z_2(x); \ x_1 = 104; \ x_2 = 291; \ x_3 = 140; \ x_4 = 155; \ x_5 = 145; \ Z_2^{0^*} = 62,395$

Max $Z_3(x); \ x_1 = 104; \ x_2 = 291; \ x_3 = 140; \ x_4 = 155; \ x_5 = 145; \ Z_3^{0^*} = 68,565$

Depending on these solutions, PIS set of (P2) are given below.

$$Z^{0^*} = \{60,475; \ 62,395; \ 68,565\}.$$

Step 4: In this step, by using $Z^{0^*} = \{60,475; \ 62,395; \ 68,565\}$, Global Criterion method and Compromise Proramming solutions are determined.

i. for $p = 1$, (P2) is organized as follows.

$$\text{Min } \left[3 - (0.003835x_1 + 0.0033611x_2 + 0.003466x_3 + 0.003621x_4 + 0.003898x_5)\right]$$

Subject to (P2.1a)

$$g_1(x): 35x_1 + 30x_2 + 38x_3 + 42x_4 + 40x_5 \leq 30,000$$

$$g_2(x): x_1 + x_2 + x_3 + x_4 + x_5 \geq 450$$

$$g_3(x): x_1 \geq 100; \ g_4(x): x_2 \geq 130$$

$$g_5(x): x_3 \geq 140; \ g_6(x): x_3 \leq 195$$

$$g_7(x): x_4 \geq 155; \ g_8(x): x_4 \leq 185$$

$$g_9(x): x_5 \geq 145; \ g_{10}(x): x_5 \leq 200$$

$$x_1, x_2, x_3, x_4, x_5 \geq 0 \text{ and integer.}$$

The variable values obtained from the solution of (P2.1a) are determined as $x_1 = 104; \ x_2 = 291; \ x_3 = 140; \ x_4 = 155; \ x_5 = 145; \ \text{Max} F = 2.988625$. According to these variable values, Min $d_1 = \left[3 - (2.988625)\right] = 0.011375$. Based on this information, $x^{01} = (104; 291; 140; 155; 145)$ and $Z^{01} = (59,785; 62,395; 68,565)$ are obtained.

ii. For $p = \infty$; (P2) is organized as follows.

$$\text{Min } d_\infty$$

Subject to (P2.1b)

$$75x_1 + 60x_2 + 80x_3 + 70x_4 + 92x_5 + 60475d_\infty \geq 60,475$$

$$89x_1 + 92x_2 + 95x_3 + 90x_4 + 0.95x_5 + 62395d_\infty \geq 62,395$$

$$90x_1 + 90x_2 + 92x_3 + 85x_4 + 93x_5 + 68565d_\infty \geq 68,565$$

$$g_1(x): 35x_1 + 30x_2 + 38x_3 + 42x_4 + 40x_5 \leq 30,000$$

$$g_2(x): x_1 + x_2 + x_3 + x_4 + x_5 \geq 450$$

$$g_3(x): x_1 \geq 100; \quad g_4(x): x_2 \geq 130$$

$$g_5(x): x_3 \geq 140; \quad g_6(x): x_3 \leq 195$$

$$g_7(x): x_4 \geq 155; \quad g_8(x): x_4 \leq 185$$

$$g_9(x): x_5 \geq 145; \quad g_{10}(x): x_5 \leq 200$$

$$x_1, x_2, x_3, x_4, x_5 \geq 0 \text{ and integer.}$$

The variable values and objective function values determined from the solution of (P2.1b) are $x^{0\infty} = (158; 228; 140; 155; 145)$ and $Z^{0\infty} = (60,055; 61,990; 68,385)$. Min $d_\infty = 0.006945$ is obtained.

Step 5: In this step, the fuzzy linear programming solution is made for each objective function. At this stage; firstly, ideal solution sets obtained from Steps 1 and 3 are used. These are $Z^{1*} = \{51,895; 52,410; 57,905\}$ and $Z^{0*} = \{60,475; 62,395; 68,565\}$. Membership functions are defined according to positive ideal solutions (Figures 7.3-7.8).
Membership function for $Z_1(x)$;

$$\mu_{\tilde{Z}_1}(x) = \begin{cases} 1 & \text{if } 60,475 < Z_1(x) \\ 1 - \dfrac{[60,475 - Z_1(x)]}{(60,475 - 51895)} & \text{if } 51,895 \leq Z_1(x) \leq 60,475 \\ 0 & \text{if } Z_1(x) < 51,895 \end{cases}$$

Membership function for $Z_2(x)$;

$$\mu_{\tilde{Z}_2}(x) = \begin{cases} 1 & \text{if } 62,395 < Z_2(x) \\ 1 - \dfrac{[62,395 - Z_2(x)]}{(62,395 - 52,410)} & \text{if } 52,410 \leq Z_2(x) \leq 62,395 \\ 0 & \text{if } Z_2(x) < 52,410 \end{cases}$$

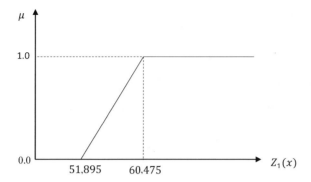

FIGURE 7.2 Fuzzy membership function for $Z_1(x)$.

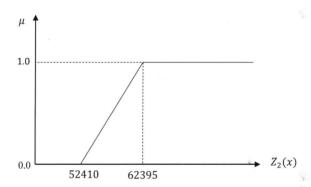

FIGURE 7.3 Fuzzy membership function for $Z_2(x)$.

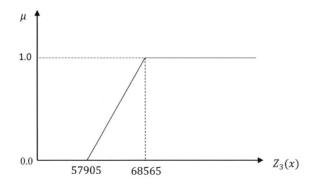

FIGURE 7.4 Fuzzy membership function for $Z_3(x)$.

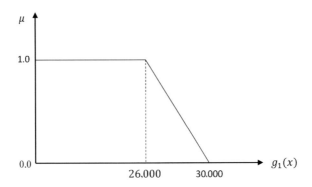

FIGURE 7.5   Membership function for budget constraint.

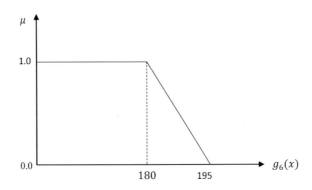

FIGURE 7.6   Fuzzy membership function for constraint $g_6(x)$.

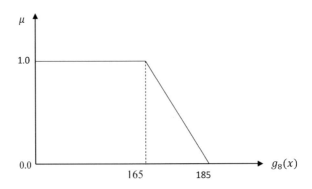

FIGURE 7.7   Fuzzy membership function for constraint $g_8(x)$.

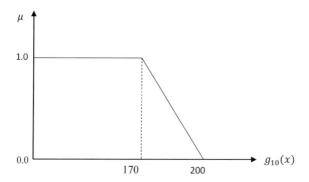

FIGURE 7.8 Fuzzy membership function for constraint $g_{10}(x)$.

Membership function for $Z_3(x)$;

$$\mu_{\tilde{Z}_3}(x) = \begin{cases} 1 & \text{if } 68,565 < Z_3(x) \\ 1 - \dfrac{[68,565 - Z_3(x)]}{(68,565 - 57,905)} & \text{if } 57,905 \le Z_3(x) \le 68,565 \\ 0 & \text{if } Z_3(x) < 57,905 \end{cases}$$

Membership functions are then created for constraint functions. Membership function for budget constraint;

$$\mu_{g_1(X)} = \begin{cases} 1 & \text{if } g_1(x) < 26,000 \\ 1 - \dfrac{[g_1(x) - 26,000]}{40,00} & \text{if } < 26,000 \le g_1(x) \le 30,000 \\ 0 & \text{if } g_1(x) > 30,000 \end{cases}$$

Membership function for $g_6(x)$;

$$\mu_{g_6(X)} = \begin{cases} 1 & \text{if } g_6(x) < 180 \\ 1 - \dfrac{[g_6(x) - 180]}{15} & \text{if } 180 \le g_6(x) \le 195 \\ 0 & \text{if } g_6(x) > 195 \end{cases}$$

Membership function for $g_8(x)$;

$$\mu_{g_8(X)} = \begin{cases} 1 & \text{if } g_8(x) < 165 \\ 1 - \dfrac{[g_8(x) - 165]}{20} & \text{if } 165 \le g_8(x) \le 185 \\ 0 & \text{if } g_8(x) > 185 \end{cases}$$

Membership function for $g_{10}(x)$;

$$\mu_{g_{10}(x)} = \begin{cases} 1 & \text{if } g_{10}(x) < 170 \\ 1 - \dfrac{[g_{10}(x) - 170]}{30} & \text{if } 170 \le g_{10}(x) \le 200 \\ 0 & \text{if } g_{10}(x) > 200 \end{cases}$$

Using fuzzy membership functions, the problem is arranged mathematically as follows.

$$75x_1 + 60x_2 + 80x_3 + 70x_4 + 92x_5 - 8580\alpha \ge 51,895$$

$$89x_1 + 92x_2 + 95x_3 + 90x_4 + 0.95x_5 - 9985\alpha \ge 52,410$$

$$90x_1 + 90x_2 + 92x_3 + 85x_4 + 93x_5 - 10,660\alpha \ge 57,905$$

Subject to (P3)

$$g_1(x): 35x_1 + 30x_2 + 38x_3 + 42x_4 + 40x_5 + 4000\alpha \le 30,000$$

$$g_2(x): x_1 + x_2 + x_3 + x_4 + x_5 \ge 450$$

$$g_3(x): x_1 \ge 100; \ g_4(x): x_2 \ge 130$$

$$g_5(x): x_3 \ge 140; \ g_6(x): x_3 + 15\alpha \le 195$$

$$g_7(x): x_4 \ge 155; \ g_8(x): x_8 + 20\alpha \le 185$$

$$g_9(x): x_5 \ge 145; \ g_{10}(x): x_5 + 30\alpha \le 200$$

$$x_1, x_2, x_3, x_4, x_5 \ge 0 \text{ and integer.}$$

The results obtained from the Fuzzy Linear Programming solution for each objective function (P3) are given below.
For Max $Z_1(x)$;

$$\alpha = 0.604895, \ x_1 = 147; x_2 = 130; x_3 = 145; \ x_4 = 172; x_5 = 145 \text{ and } Z_1^* = 55,195$$

For Max $Z_2(x)$;

$$\alpha = 0.5, \ x_1 = 100; x_2 = 229; x_3 = 140; \ x_4 = 155; x_5 = 145 \text{ and } Z_2^* = 57,425$$

For Max $Z_3(x)$;

$\alpha = 0.5$, $x_1 = 100; x_2 = 229; x_3 = 140; x_4 = 155; x_5 = 145$ and $Z_3^* = 63,245$

$$Z^* = \{55,195; \ 57,425; \ 63,245\}.$$

Based on these results; $\alpha^* = \max(0.604895; 0.5; 0.5) = 0.604895$ is chosen.
Step 6: Final compromise solution steps for $\alpha^* = 0.604895$ are given below. Firstly (P1), if the membership functions in step 5 are arranged for α^*-cut $= 0.604895$, the mathematical model named (P4) is obtained.

$$\text{Max } Z_1(x): 75x_1 + 60x_2 + 80x_3 + 70x_4 + 92x_5$$

$$\text{Max } Z_2(x): 89x_1 + 92x_2 + 95x_3 + 90x_4 + 0.95x_5$$

$$\text{Max } Z_3(x): 90x_1 + 90x_2 + 92x_3 + 85x_4 + 93x_5$$

Subject to (P4)

$$g_1(x): 35x_1 + 30x_2 + 38x_3 + 42x_4 + 40x_5 \leq 27580.42$$

$$g_2(x): x_1 + x_2 + x_3 + x_4 + x_5 \geq 450$$

$$g_3(x): x_1 \geq 100; \ g_4(x): x_2 \geq 130$$

$$g_5(x): x_3 \geq 140; \ g_6(x): x_3 \leq 185.926575$$

$$g_7(x): x_4 \geq 155; \ g_8(x): x_4 \leq 172.9021$$

$$g_9(x): x_5 \geq 145; \ g_{10}(x): x_5 \leq 181.85315$$

$$x_1, x_2, x_3, x_4, x_5 \geq 0 \text{ and integer.}$$

Positive ideal solutions for (P4) are determined at first.
Max $Z_1(x)$ için;

$$x_1 = 100; \ x_2 = 215; \ x_3 = 140; \ x_4 = 155; \ x_5 = 145 \text{ and } Z_1^* = 54,925$$

Max $Z_2(x)$ için;

$$x_1 = 100; \ x_2 = 215; \ x_3 = 140; \ x_4 = 155; \ x_5 = 145 \text{ and } Z_2^* = 56,375$$

Max $Z_3(x)$ için;

$$x_1 = 100; \ x_2 = 215; \ x_3 = 140; \ x_4 = 155; \ x_5 = 145 \text{ and } Z_3^* = 62,125$$

Based on these results, positive ideal solution set of (P4) is {54925; 56375; 62125}. By using positive ideal solution set;

i. For $p = 1$;

$$\text{Min } G : \left[3 - \left(0.004233x_1 + 0.0037111x_2 + 0.003825x_3 \right. \right.$$
$$\left. \left. + 0.003996x_4 + 0.004303x_5 \right) \right]$$

Max $0.004233x_1 + 0.0037111x_2 + 0.003825x_3 + 0.003996x_4 + 0.004303x_5$

Subject to (4.1a)

$$g_1(x): 35x_1 + 30x_2 + 38x_3 + 42x_4 + 40x_5 \leq 27580.42$$

$$g_2(x): x_1 + x_2 + x_3 + x_4 + x_5 \geq 450$$

$$g_3(x): x_1 \geq 100; \; g_4(x): x_2 \geq 130$$

$$g_5(x): x_3 \geq 140; \; g_6(x): x_3 \leq 185.926575$$

$$g_7(x): x_4 \geq 155; \; g_8(x): x_4 \leq 172.9021$$

$$g_9(x): x_5 \geq 145; \; g_{10}(x): x_5 \leq 181.85315$$

$$x_1, x_2, x_3, x_4, x_5 \geq 0 \text{ and integer.}$$

From the solution of (P4.1a), $x_1 = 100$; $x_2 = 215$; $x_3 = 140$; $x_4 = 155$ ve $x_5 = 145$ variable values are obtained. According to this, each objective function is obtained as $Z_1^1 = 54,925$; $Z_2^1 = 56,375$; $Z_3^1 = 62,125$. Min $G:(3-3) = 0$ indicates that all objectives realize in their own positive ideal solution.

ii. For $p = \infty$

$$\text{Min } d_\infty$$

Subject to (4.1b)

$$75x_1 + 60x_2 + 80x_3 + 70x_4 + 92x_5 + 54,925d_\infty \geq 54,295$$

$$89x_1 + 92x_2 + 95x_3 + 90x_4 + 0.95x_5 + 56,375d_\infty \geq 56,375$$

$$90x_1 + 90x_2 + 92x_3 + 85x_4 + 93x_5 + 62,125d_\infty \geq 62,125$$

$$g_1(x): 35x_1 + 30x_2 + 38x_3 + 42x_4 + 40x_5 \leq 27580.42$$

$$g_2(x): x_1 + x_2 + x_3 + x_4 + x_5 \geq 450$$

$$g_3(x): x_1 \geq 100; \; g_4(x): x_2 \geq 130$$

$$g_5(x): x_3 \geq 140; \; g_6(x): x_3 \leq 185.926575$$

$$g_7(x): x_4 \geq 155; \; g_8(x): x_4 \leq 172.9021$$

$$g_9(x): x_5 \geq 145; \; g_{10}(x): x_5 \leq 181.85315$$

$$x_1, x_2, x_3, x_4, x_5 \geq 0 \text{ and integer.}$$

The variable values obtained from the solution of (4.1b) are $x_1 = 100$; $x_2 = 215$; $x_3 = 140$; $x_4 = 155$; $x_5 = 145$. On the other hand, each objective function value is obtained as $Z_1 = 54,925$; $Z_2 = 56,375$; $Z_3 = 62,125$. Each objective function value of (P4.1b) is designated as $d_\infty = 0$. Thereby, each objective has realized in its own positive ideal solution.

With the Fuzzy GCM algorithm consisting of six steps, variable values and goal function values obtained in each step are given in Table 7.4.

It is possible to analyze the results in Table 7.4 by dividing them into three groups.

i. In Step-1, a positive ideal solution set for (P1) has been designated. These values indicate the best performance of each objective function. Due to the fact that objective function values realize in different values of the variables, there is no feasible solution in this step. Actually, non-dominated solutions can be determined using these information. By using the information in Step-1, non-dominated solutions table is given in Table 7.5.

The values in Table 7.5 include the results that Global Criterion Method and CP will give in terms of nondominated solution. This situation is

TABLE 7.4
Fuzzy Global Criterion Method Results

Step-1	$Z^{1*} = \{51,895; \; 52,410; \; 57,905\}$	No feasible solution
Step-2	$Z^{11} = (51,775; 52,410; 57,905)$	$x^{11} = (102; \; 160; \; 140; \; 155; \; 146)$
	$Z^{1\infty} = (51,835; 52,320; 57,865)$	$x^{1\infty} = (114; \; 146; \; 140, \; 155; \; 145)$
Step-3	$Z^{0*} = \{60,475; \; 62,395; \; 68,565\}$	No feasible solution
Step-4	$Z^{01} = \{59,785; \; 62,395; \; 68,565\}$	$x^{01} = (104; \; 291; \; 140; \; 155; \; 145)$
	$Z^{0\infty} = (60,055; 61,990; \; 68,385)$	$x^{0\infty} = (158; \; 228; \; 140; \; 155; \; 145)$
Step-5	$Z^* = \{55,195; \; 57,425; \; 63,245\}$	Fuzzy LP Solutions
Step-6	$Z^* = \{54,925; \; 56,375; \; 62,125\}$	$x^{11} = (100; \; 215; \; 140; \; 155; \; 145)$
	$Z^1 = \{54,925; \; 56,375; \; 62,125\}$	$x^1 = (100; \; 215; \; 140; \; 155; \; 145)$
	$Z^\infty = \{54,925; \; 56,375; \; 62,125\}$	$x^\infty = (100; \; 215; \; 140; \; 155; \; 145)$

TABLE 7.5

Non-Dominated Solution of (P1)

	Z_1	Z_2	Z_3
x^{1*}	51,895	52,230	57,825
x^{2*}	51,775	52,410	57,905
x^{3*}	51,775	52,410	57,905

explainable by using the information in Steps-1 and 2. Firstly, (%) value of each objective function has been calculated based on the results in Step-1.

$$Z_1 = \frac{51895}{698} = 74.34814$$

$$Z_2 = \frac{52410}{702} = 74.658125$$

$$Z_3 = \frac{57905}{702} = 82.48575$$

These rates show the average % value of the purchased material. In other words, they show the average performance value of each objective. The calculations can be made for Table 7.5 (%) transformation of non-dominated solutions are given in Table 7.6.

Ranges of each objective function from the results in Table 7.6 that realizes in terms of Global Criterion Method and CP are determined as $73.75356 \leq Z_1 \leq 74.34814$; $74.65812 \leq Z_2 \leq 74.82808$; $82.48575 \leq Z_2 \leq 82.84384$.

From the information from Step-2,

for $p = 1$ (Global Criterion Method),

$$Z_1 = \frac{51,775}{702} = 7,375,356; \ Z_2 = \frac{52,410}{702} = 74.658125;$$

$$Z_3 = \frac{57,905}{702} = 82.48575$$

TABLE 7.6

Nondominated Solutions in Terms of Percentage

	Z_1	Z_2	Z_3
x^{1*}	74.34814	74.82808	82.84384
x^{2*}	73.75356	74.658125	82.48575
x^{3*}	73.75356	74.65812	82.48575

These results have realized in lower limits of % ranges of non-dominated solutions.

For $p = \infty$ (CP),

$$Z_1 = \frac{51,835}{700} = 74.05; \ Z_2 = \frac{52,320}{700} = 7,474,285,714;$$

$$Z_3 = \frac{57,865}{700} = 8,266,428,571$$

The % values obtained in terms of CP has been within the % ranges of the non-dominated solutions and higher than the lower limits. Two solution results are an expected situation in terms of distance family. The solution obtained for $p = \infty$ is dominate the results given by $p = 1$.

ii. The difference in these steps stems from increasing right hand side values of $g_1(x)$, $g_6(x)$, $g_8(x)$ and $g_{10}(x)$ constraints in accordance with the information coming from the decision maker. The values designated in these steps are given in Table 7.4. It is possible to make similar calculations and comments given above for Steps 3 and 4.

iii. In Steps 5 and 6, the decision is made according to the positive ideal solutions obtained from Steps 1 and 3. From the fuzzy linear programming perspective in Step-5, firstly, membership functions were created for each goal function using $Z^{1*} = \{51895; \ 52410; \ 57905\}$ and $Z^{0*} = \{60,475; \ 62,395; \ 68,565\}$ afterwards, fuzzy model was established by defining membership functions for constraint functions. In Step-5, the established model is solved for each objective function, and the most satisfactory one is preferred among these solutions. In Step-6, firstly, by using α^*-cut determined from $\alpha^* = \max_k \alpha_k^*$, right hand side values of $g_1(x)$, $g_6(x)$, $g_8(x)$ and $g_{10}(x)$ constraint functions have been assigned.

In (P4), each goal function is realized at the same value of same decision variables. According to Lai and Hwang (1994), this solution type in MODM problems are called optimal solution. Therefore, the solutions obtained from (P4.1a) and (P4.1b) are optimal solutions. When these results are taken into consideration, it is seen that positive ideal solution set and compromised solutions are equal to each other in the problem. Objective function values for these three situations have been calculated below as %.

$$Z_1 = \frac{54,925}{755} = 7,274,834; Z_2 = \frac{56,375}{755} = 7,466,887; Z_3 = \frac{62,125}{755} = 82.48575$$

Accordingly, when 100, 215, 140, 155 and 145 material are purchased respectively, on-time delivery performance mean of products are calculated as 72.74834%, green product quality mean is 74.66887%, and Work Safety and Labor Health average performance of the suppliers are 82.48575%.

7.5 RESULT AND SUGGESTIONS

In supplier selection evaluation and selection process, the criteria designated based on the preference of decision maker and the method to be used directly affect the result. This is because how to select the supplier through the criteria determined from economic, environmental and social dimension is important. For this reason, experience and knowledge of the decision-maker plays a critical role in this process. With this point of view, such kind of a results appears: in creating sustainability, not only the performance of suppliers, but also the criteria that decision-maker will assign and method(s) preferred are effective. In addition to them, many factors like policy of the business, geographical location, global warming, developmental level of the country are important factors affecting sustainability. For sure, it is impossible to ensure sustainability with (%) senchron of these factors. However, it is possible to approach sustainability by minimizing sustainability.

In this study, supplier selection problem of a business operating in construction sector have been handled. The business has planned purchasing heat and noise insulation material in different amounts based on three different criteria determined from current five suppliers. In the previous periods, the company has made its buying plan according to the total cost, quality, transportation cost and technical capability criteria, especially due to the Covid-19 outbreak, the business wanted to create a new purchasing plan according to 'On-Time Delivery Performance Level of the Product', 'Green Product Quality' and 'Work Safety and Labor Health' criteria. This is because Covid-19 outbreak has caused delays in delivery of the products, and in order to overcome this negativity, on-time delivery level of each supplier has been determined from the previous period information. As heat and noise insulation materials have chemical ingredient, Green Product Quality and Labor Health criteria are significant for the position in the sector of the business. When this buying plan of the business are considered, the business of the criteria has changed. In this study, Fuzzy GCM algorithm has been used for solving S-SSP of the business. The most important feature of this algorithm is that it includes three different methods. By using Global Criterion Method and CP that are used in algorithm, the farthest and closest values to the positive ideal solution set have been determined. Afterward, α-cut level that satisfies objectives at maximum level have been designated using Fuzzy Linear Programming. Besides, results of these three methods are the same in the 6th step of the algorithm. This stems from that objective functions are in maximization type and stems from budget constraint.

In methods that are in the algorithm suggested by Leung (1984), weight of objective functions has not been used. This algorithm can be developed by adding weight factors for objectives to the algorithm. Although weight factors are not used in Global Criterion Method, the algorithm can be modified by using 'Weighted Global Criterion Method' suggested by Marler and Arora (2004). In assigning weights, by using the Analytic Hierarchy Process, the decision maker can be included in the algorithm.

ACKNOWLEDGMENT

I would like to express my gratitude to Akyamaç Engineering & Construction Firm and its manager Emrah Gür for their valuable cooperation. Besides, I would like to give my special thanks to Tandoğan Türkmen Reis who has encouraged me during my academic life.

REFERENCES

Acar, M. F., & Çapkın, A. (2017). Analitik ağ süreci ile tedarikçi seçimi: Otomotiv sektörü örneği. *Adnan Menderes Üniversitesi Sosyal Bilimler Enstitüsü Dergisi, 4*(2), 121–134.

Achillas, C., Bochtis, D., Aidonis, D., Folinas, D. (2019). *Green supply chain management.* London: Routledge, https://doi.org/10.4324/9781315628691.

Altuntaş, C., & Türker, D. (2012). Sürdürülebilir tedarik zincirleri: Sürdürülebilirlik raporlarının içerik analizi. *Dokuz Eylül Üniversitesi Sosyal Bilimler Enstitüsü Dergisi, 14*(3), 39–64.

Amid, A., Ghodsypour, S. H., & O'Brien, C. (2011). A weighted max–min model for fuzzy multi-objective supplier selection in a supply chain. *International Journal of Production Economics, 131*(1), 139–145.

Aouadni, S., Allouche, M. M., & Rebaï, A. (2013). Supplier selection: An analytic network process and imprecise goal programming model integrating the decision-maker's preferences. *International Journal of Operational Research, 6*(2), 137–154.

Arikan, F. (2013). A fuzzy solution approach for multi objective supplier selection. *Expert Systems with Applications, 40*(3), 947–952.

Ashlaghi M. J. (2014). A new approach to green supplier selection based on fuzzy multi-criteria decision making method and linear physical programming. *Tehnički Vjesnik, 21*(3), 591–597.

Ayhan, M. B., & Kilic, H. S. (2015). A two stage approach for supplier selection problem in multi-item/multi-supplier environment with quantity discounts. *Computers & Industrial Engineering, 85*, 1–12.

Azadnia, A. H., Saman, M. Z. M., & Wong, K. Y. (2014). Sustainable supplier selection and order lot-sizing: An integrated multi-objective decision-making process. *International Journal of Production Research, 53*(2), 383–408.

Bache, J., Carr, R., Parnaby, J., & Tobias, A. M. (1987). Supplier development systems. *International Journal of Technology Management, 2*(2), 219–228.

Bai, C., & Sarkis, J. (2010). Integrating sustainability into supplier selection with grey system and rough set methodologies. *International Journal of Production Economics, 124*(1), 252–264.

Banasik, A., Bloemhof-Ruwaard, J. M., Kanellopoulos, A., Claassen, G. D. H., & van der Vorst, J. G. A. J. (2018). Multi-criteria decision making approaches for green supply chains: A review. *Flexible Services and Manufacturing Journal, 30*(3), 366–396. doi:10.100710696-016-9263-5

Chai, J., Liu, J. N. K., & Ngai, E. W. T. (2013). Application of decision-making techniques in supplier selection: A systematic review of literature. *Expert Systems with Applications, 40*(10), 3872–3885.

Chai, J., & Ngai, E. W. (2019). Decision-making techniques in supplier selection: Recent accomplishments and what lies ahead. *Expert Systems with Applications, 140*, 112903.

Chang, N. B., & Lu, H. Y. (1997). A new approach for long term planning of solid waste management systems using fuzzy global criterion. *Journal of Environmental Science & Health Part A, 32*(4), 1025–1047.

Chang, N. B., & Pires, A. (2015). *Sustainable solid waste management: A systems engineering approach.* New Jersey: John Wiley & Sons.

Chen, W., Mei, H., Chou, S. Y., Luu, Q. D., & Yu, T. H. K. (2016). A fuzzy MCDM approach for green supplier selection from the economic and environmental aspects. *Mathematical Problems in Engineering*, Vol. 2016, 1–10. http://dx.doi.org/10.1155/2016/8097386

Cheraghi, S. H., Dadashzadeh, M., & Subramanian, M. (2004). Critical success factors for supplier selection: An update. *Journal of Applied Business Research*, *20*(2), 91–108.

Chiou, C. Y., Hsu, C. W., & Hwang, W. Y. (2008). Comparative investigation on green supplier selection of the American, Japanese and Taiwanese electronics industry in China. *In 2008 IEEE International Conference on Industrial Engineering and Engineering Management*, 1909–1914.

Cohon, J. (1978). *Multiobjective programming and planning.* New York: Academic Press.

Choudhary, D., & Shankar, R. (2014). A goal programming model for joint decision making of inventory lot-size, supplier selection and carrier selection. *Computers & Industrial Engineering*, *71*, 1–9.

Çakın, E., & Özdemir, A. (2013). Using analytic network process and electr e methods for supplier selection decision and an application. *Journal of Economics & Administrative Sciences*, *15*(2), 339–364.

Çalık, A. (2018). Comparison of fuzzy multi-objective linear programming approaches in green supplier selection. *Turkish Studies Information Technologies & Applied Sciences*, *13*(13), 1–21.

De Boer, L., Labro, E., & Morlacchi, P. (2001). A review of methods supporting supplier selection. *European Journal of Purchasing and Supply Management*, 7-2, 75–89.

Dempsey W. A. (1978).Vendor selection and the buying process. *Industrial Marketing Management*, *7*, 257–267.

Dickson G. W. (1966). An analysis of vendor selection systems and decisions. *Journal of Purchasing*, *1*, 5–17.

Ding, H., Benyoucef, L., & Xie, X. (2005). A simulation optimization methodology for supplier selection problem. *International Journal of Computer Integrated Manufacturing*, 18(2-3), 210–224.

Ekhtiari M., Zandieh M., Alem-Tabriz, A., & Rabieh M. (2018). A nadir compromise programming for supplier selection problem under uncertainty. *IJIEPR*, *29*(1), 1–14.

Elahi, B., Seyed-Hosseini, S. M., & Makui, A. (2011). A fuzzy compromise programming solution for supplier selection in quantity discounts situation. *International Journal of Industrial Engineering & Production Research*, *22*(2), 107–114.

Gaballa, A. A. (1974). Minimum cost allocation of tenders. *Operational Research Quarterly*, *25*(3), 389–398.

Genovese, A., Koh, S. L., Bruno, G., & Bruno, P. (2010). Green supplier selection: A literature review and a critical perspective. In *2010 8th International Conference on Supply Chain Management and Information*, 1–6.

Ghodsypour, S. H., & O'Brien, C. (1998). A decision support system for supplier selection using an integrated analytic hierarchy process and linear programming. *International Journal of Production Economics*, *56–57*, 199–212. doi:10.1016/S0925-5273(97)00009-1

Grover, R., Grover, R., Rao, V. B., & Kejriwal, K. (2016). Supplier selection using sustainable criteria in sustainable supply chain management. *International Journal of Economics and Management Engineering*, *10*, 1775–1780.

Ghoushchi, S. J., Milan, M. D., & Rezaee, M. J. (2018). Evaluation and selection of sustainable suppliers in supply chain using new GP-DEA model with imprecise data. *Journal of Industrial Engineering International*, *14*(3), 613–625.

Güçlü, P., & Özdemir, A. (2015). Bulanık hedef programlama ile tedarik zinciri optimizasyonu: Tekstil sektöründe bir uygulama. *Hacettepe Üniversitesi İktisadi ve İdari Bilimler Fakültesi Dergisi*, *33*(1), 79–100.

Govindan, K., Rajendran, S., Sarkis, J., & Murugesan, P. (2015). Multi criteria decision making approaches for green supplier evaluation and selection: A literature review. *Journal of Cleaner Production*, *98*, 66–83.

Haeri, S. A. S., & Rezaei, J. (2019). A grey-based green supplier selection model for uncertain environments. *Journal of Cleaner Production*, *221*, 768–784.

Ho, H.-P. (2019). The supplier selection problem of a manufacturing company using the weighted multi-choice goal programming and minmax multi-choice goal programming. *Applied Mathematical Modelling*, *75*, 819–836.

Hu, K.-J., & Yu, V. F. (2016). An integrated approach for the electronic contract manufacturer selection problem. *Omega*, *62*, 68–81.

Humphreys, P. K., Wong, Y. K., & Chan, F. T. S. (2003). Integrating environmental criteria into the supplier selection process. *Journal of Materials Processing Technology*, *138*(1–3), 349–356.

Jadidi, O., Cavalieri, S., & Zolfaghari, S. (2015). An improved multi-choice goal programming approach for supplier selection problems. *Applied Mathematical Modelling*, *39*, 4213–4222.

Kaviani, M. A., Peykam, A., Khan, S. A., Brahimi, N., & Niknam, R. (2020). A new weighted fuzzy programming model for supplier selection and order allocation in the food industry. *Journal of Modelling in Management*, *15*(2), 381–406.

Kazemi, N., Ehsani, E., & Glock, C. H. (2014). Multi-objective supplier selection and order allocation under quantity discounts with fuzzy goals and fuzzy constraints. *International Journal of Applied Decision Sciences*, *7*(1), 66–96.

Khan, S. A., Kusi-Sarpong, S., Arhin, F. K., & Kusi-Sarpong, H. (2018). Supplier sustainability performance evaluation and selection: A framework and methodology. *Journal of Cleaner Production*, *205*, 964–979.

Ku, C., Chang, C., & Ho, H. (2010). Global supplier selection using fuzzy analytic hierarchy process and fuzzy goal programming. *Quality and Quantity*, *44*, 623–640.

Lai, Y. J., & Hwang, C. L. (1994). *Fuzzy multiple objective decision making: Methods and applications*. Berlin: Springer-Verlag.

Laínez-Aguirre, J. M., & Puigjaner, M. (2015). *Advances in integrated and sustainable supply chain planning*. Switzerland: Springer.

Leung, Y. (1984). Compromise programming under fuzziness. *Control and Cybernetics*, *13*, 203–215.

Li, J., Fang, H., & Song, W. (2019). Sustainable supplier selection based on SSCM practices: A rough cloud TOPSIS approach. *Journal of Cleaner Production*, *222*, 606–621.

Liao, C. N., & Kao, H. P. (2010). Supplier selection model using taguchi loss function, analytical hierarchy process and multi-choice goal programming. *Computers & Industrial Engineering*, *58*(4), 571–577.

Liu, S., & Papageorgiou, L. G. (2013). Multiobjective optimisation of production, distribution and capacity planning of global supply chains in the process industry. *Omega*, *41*(2), 369–382.

Lo, H-W., Liou, J. J. H., Wang, H-S., & Tsai, Y-S. (2018). An integrated model for solving problems in green supplier selection and order allocation. *Journal of Cleaner Production*, *190*, 339–352.

Marler, R. T., & Arora, J. S. (2004). Survey of multi-objective optimization methods for engineering. *Structural and Multidisciplinary Optimization*, *26*(6), 369–395.

Miettinen, K. (2012). *Nonlinear multiobjective optimization* (Vol. 12). New York: Springer Science & Business Media.

Mirzaee, H., Naderi, B., & Pasandideh, S. H. R. (2018). A preemptive fuzzy goal programming model for generalized supplier selection and order allocation with incremental discount. *Computers & Industrial Engineering*, *122*, 292–302.

Moghaddam, K. S. (2015). Supplier selection and order allocation in closed-loop supply chain systems using hybrid Monte Carlo simulation and goal programming. *International Journal of Production Research*, *53*(20), 6320–6338.

Mohammed, A., Harris, I., & Govindan, K. (2019). A hybrid MCDM-FMOO approach for sustainable supplier selection and order allocation. *International Journal of Production Economics*, *217*, 171–184.

Moheb-Alizadeh, H., & Handfield, R. (2019). Sustainable supplier selection and order allocation: A novel multi-objective programming model with a hybrid solution approach. *Computers & Industrial Engineering*, *129*, 192–209.

Moore, D. L., & Fearon, H. E. (1973). Computer-assisted decision-making in purchasing. *Journal of Purchasing*, *9*(4), 5–25.

Mukherjee, K. (2017). *Supplier selection an MCDA-based approach*. India: Springer. doi:10.1007/978-81-322-3700-6

Nazari-Shirkouhi, S., Shakouri, H., Javadi, B., & Keramati, A. (2013). Supplier selection and order allocation problem using a two-phase fuzzy multi-objective linear programming. *Applied Mathematical Modelling*, *37*(22), 9308–9323.

Osman, H., & Demirli, K. (2010). A bilinear goal programming model and a modified benders decomposition algorithm for supply chain reconfiguration and supplier selection. *International Journal of Production Economic*, *124*(1), 97–105.

Ozkok, B. A., & Tiryaki, F. (2011). A compensatory fuzzy approach to multi-objective linear supplier selection problem with multiple-item. *Expert Systems with Applications*, *38*(9), 11363–11368.

Özçelik F., & Avcı Öztürk, B. A. (2014). Research on barriers to sustainable supply chain management and sustainable supplier selection crıteria. *Dokuz Eylül Üniversitesi Sosyal Bilimler Enstitüsü Dergisi*, *16-2*, 259–279.

Ozyoruk, B. (2018). A literature survey on green supplier selection, *The Eurasia Proceedings of Science, Technology [EPSTEM] Engineering & Mathematics*, *2*, 407–411.

Pandey, P., Shah, B. J., & Gajjar, H. (2017). A fuzzy goal programming approach for selecting sustainable suppliers. *Benchmarking: An International Journal*, *24*(5), 1138–1165.

Rabieh, M., Babaee, L., Fadaei Rafsanjani, A., & Esmaeili, M. (2019). Sustainable supplier selection and order allocation: An integrated delphi method, fuzzy topsİs and multi-objective programming model. *Scientia Iranica*, *26*(4), 2524–2540.

Roa C. P., & Kiser, G. E. (1980). Educational buyer's perception of vendor attributes. *Journal of Purchasing Materials Management*, *16*, 25–30.

Sağnak, M. (2020). Mobilya sektöründe sürdürülebilir tedarik zinciri yönetimi performans değerlendirmesi. *Uluslararası İktisadi ve İdari İncelemeler Dergisi*, *26*, 97–114.

Shaw, K., Shankar, R., Yadav, S. S., & Thakur, L. S. (2012). Supplier selection using fuzzy AHP and fuzzy multi-objective linear programming for developing low carbon supply chain. *Expert Systems with Applications*, *39*(9), 8182–8192.

Sheikhalishahi, M., & Torabi, S. A. (2014). Maintenance supplier selection considering life cycle costs and risks: A fuzzy goal programming approach. *International Journal of Production Research*, *52*(23), 7084–7099.

Shih, C. J., & Chang, C. J. (1995). Pareto optimization of alternative global criterion method for fuzzy structural design. *Computers & Structures*, *54*(3), 455–460.

Sivrikaya, B. T., Kaya, A., Dursun, M., & Çebi, F. (2015). Fuzzy AHP–goal programming approach for a supplier selection problem. *Research in Logistics & Production*, *5*(3), 271–285.

Seuring, S., & Müller, M. (2008). From a literature review to a conceptual framework for sustainable supply chain management. *Journal of Cleaner Production*, *16*(15), 1699–1710.

Seçkin, F. (2019). Tedarik zinciri yönetiminde ve tedarikçi seçiminde sürdürülebilirlik kavramının gelişimi. *AURUM Journal of Engineering Systems and Architecture*, *2*(2), 45–64.

Şenocak, A. A., & Gören, H. G. (2016). An İntegrated approach for sustainable supplier selection in fuzzy environment. *International Journal of Intelligent Systems and Applications in Engineering*, 4(Special Issue-1), 130–135.

Tabucanon, M. T. (1988). *Multiple criteria decision making in industry* (Vol. 8). New York: Elsevier Science Ltd.

Thiruchelvam, S., & Tookey, J. E. (2011). Evolving trends of supplier selection criteria and methods. *International Journal of Automotive and Mechanical Engineering*, 4, 437–454.

Trisna, T., Marimin, M., Arkeman, Y., & Sunarti, T. C. (2016). Multi-objective optimization for supply chain management problem: A literature review. *Decision Science Letters*, 5(2), 283–316.

Umarusman, N., & Hacıvelioğulları, T. (2018). Solution proposal for supplier selection problem: An application İn agricultural machinery sector with global criterion method. *Dokuz Eylül Üniversitesi İktisadi ve İdari Bilimler Fakültesi Dergisi*, 33(1), 353–368.

Umarusman, N. (2019). Multiple objective decision-making methods in supplier selection. In S. K. Mangla, S. Luthra, S. K. Jakhar, A. Kumar, & N. P. Rana (Eds.), *Sustainable Procurement in Supply Chain Operations* (pp. 177–208). Boca Raton, FL: CRC Press. doi:10.1201/9780429466328-8.

Umarusman, N. & Türkmen, A. (2013), Building optimum production settings using De Novo programming with global criterion method. *International Journal of Computer Applications (0975 – 8887)*, 82(18), 49–57.

Umarusman, N., & Hacıvelioğulları, T. (2020). Compromise optimal system design for solving multi-objective green supplier selection problems. In Syed Abdul Rehman Khan (Ed.), *Global Perspectives on Green Business Administration and Sustainable Supply Chain Management* (pp. 241–275). Hershey PA, USA: IGI Global.

Ware, N. R., Singh, S. P., & Banwet, D. K. (2012). Supplier selection problem: A state-of-the-art review. *Management Science Letters*, 2(5), 1465–1490.

Weber C. A., Current, J. R., & Benton W. C. (1991). Vendor selection criteria and methods. *European Journal of Operational Research*, 50, 1–17.

Weber, C. A., & Ellram, L. M. (1993). Supplier selection using multi-objective programming: A decision support system approach. *International Journal of Physical Distribution & Logistics Management*, 23(2), 3–14.

Yousefi, S., Mahmoudzadeh, H., & Jahangoshai Rezaee, M. (2017). Using supply chain visibility and cost for supplier selection: A mathematical model. *International Journal of Management Science and Engineering Management*, 12(3), 196–205.

Yu, P. L. (1973). A class of solutions for group decision problems. *Management Science*, 19-8, 936–946.

Yücel, A., & Güneri, A. F. (2011). A weighted additive fuzzy programming approach for multi-criteria supplier selection. *Expert Systems with Applications*, 38(5), 6281–6286.

Zeleny, M. (1973). Compromise programming. In J. L. Cochrane, & M. Zeleny (Eds.), *Multiple Criteria Decision Making* (pp. 262–301). Columbia, SC: University of South Carolina.

8 Sustainability and OEE Gains in Manufacturing Operations Through TPM

Mukesh Kumar
Department of Mechanical Engineering,
National Institute of Technology, Patna, Bihar, India

Rahul S. Mor
Department of Food Engineering, National
Institute of Food Technology Entrepreneurship
and Management, Sonepat, Haryana, India

Sarbjit Singh
Department of Industrial & Production Engineering,
National Institute of Technology, Jalandhar, Punjab, India

Vikas Kumar Choubey
Department of Mechanical Engineering,
National Institute of Technology, Patna, Bihar, India

CONTENTS

8.1 INTRODUCTION

Total productive maintenance (TPM) in a manufacturing industry plays a vital role in enhanced competitiveness and to achieve a minimum breakdown of machinery and high-quality products. TPM is the Japanese innovative concept that is traced back to 1951 when preventive maintenance was introduced in Japan. Nippondenso was the first industry to implement preventive maintenance technique and thereafter they combine autonomous maintenance and preventive maintenance to achieve high success (Mor et al., 2019). Brah and Chong (2004), Agustiady and Cudney (2018) stated that TPM builds a close relationship between maintenance and productivity that shows how good care and maintenance of equipment are done. Ahuja and Khamba (2009), Samadhiya and Agrawal (2020) depict that TPM leads industry towards growth and sustainability by lowering the maintenance cost and increasing efficiency as well as increasing available time for production. Sahoo (2020) explored the utilization of TQM and maintenance and found the best way to improve the sustainability of Indian manufacturing industries. Maalem et al. (2020), Adhiutama et al. (2020) applied TPM to increase the business sustainability of the industry. Borris (2006) mentioned that TPM as a manufacturing strategy is highly competitive against a series of new rivals for years. Another approach of TPM is the reliability centered maintenance and these methods started their lives in America. In the United State, it was known as PM (no 'T') i.e. Preventive Maintenance (Nakajima, 1988; Bhardwaj et al., 2018). Three words of TPM indicates that:

- Total: Total means involvement of all employees of the system from the top management level to shop floor level, which gives strong coordination between high authority levels to a low-level employee of the system.
- Productive: Productive means there is no activity, which produces any loss in the production of goods and services that meet or fulfill customer requirements.
- Maintenance: To keep all the equipment of industry in good working condition. This aim to achieve maximum availability of the machine, high-quality product and zero breakdowns of machines at low maintenance cost.

Huang et al. (2003) conclude that the industries should be a part of the qualitative competition to improve and optimize their productivity. It will lead to improved skilled employees and efficient machines. Nakajima (1988) introduced overall equipment effectiveness (OEE) as a key performance indicator for the performance measure of TPM. Solke and Singh (2018) concluded that TPM helps in the effective utilization of resources and continuous product improvement in the industry. Mwanza and Mbohwa (2015), Chaudhuri and Jayaram (2019), Chen et al. (2019) stated that TPM implementation reduces in-process defects and losses that will help to improve business sustainability of the industry as well as lead to increase in profit of company and competitiveness. Ibrahim et al. (2019) stated that the industry moves towards sustainability by improving its maintenance schedule and reducing losses. Hami et al. (2019), Gupta et al. (2015) developed a framework and establish a relationship between sustainable maintenance and manufacturing and social sustainability. Many other authors implemented TPM in the auto part industry for improving productivity and applied TPM in the garment industry to improve OEE. Another key problem in decreasing the lot sizing is higher die setup time and it must be minimum possible for producing at competitive costs (Kumar *et al.*, 2017; Kigsirisin *et al.*, 2016; Ahmad *et al.*, 2018; Mor *et al.*, 2018; Wudhikarn, 2012).

For better organization, this chapter is divided into five sections, Section 1 is the introductory part. Section 2 discusses the problem formulation, scope of the study and methodology. Section 3 is the analysis and implementation of TPM and single minute exchange of die (SMED) along with cause–effect analysis. Section 4 is the results and discussion, the benefits ensued from TPM, SMED implementation, whereas the final Section 5 concludes the undertaken research and future research directions in the area.

8.2 PROBLEM FORMULATION

A questionnaire is developed by visiting the industry and assessing the current situation. All the relevant data of the industry is obtained through the scheduled questionnaire based on current working conditions. The procedure is mentioned below (Figure 8.1).

Following methods are used to collect data:

- Questionnaire survey,
- Interview with the employee,
- Company records,
- Breakdown summary sheet

The sampling test is used to minimize the error in data collection through a set of questions. A 25% sampling intensity has been used for minimum sample error, as suggested by Gupta and Vardhan (2016). The responses from 60 employees are considered as a sample out of 180 employees in the industry.

$$\frac{n}{N} \times 100 = C \tag{1}$$

$$N = 180 \times 30/100 = 60$$

Questions are asked in the questionnaire from all TPM pillars assuming that all the pillars are of equal importance towards the successful implementation of TPM. A breakdown data of each equipment has been collected through a breakdown summary sheet for one month period available in the industry. The failure causes of each machine

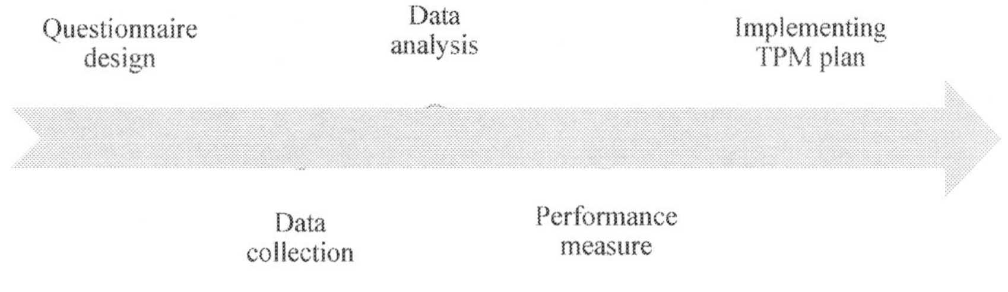

FIGURE 8.1 Research methodology.

are identified and analyzed with the help of the Pareto diagram and the cause–effect diagram. Further, the data regarding six big losses in the industries at the time of production is collected through 'an OEE/loss summary' sheet available in the industry. OEE is calculated for all machines as a performance measure. OEE of machines is calculated two times i.e. before and after TPM implementation on selected machines. Machines having minimum availability and lesser performance rate are analyzed through the implementation of 5'S, autonomous maintenance and TPM pillars.

The collected data is analyzed using a Pareto chart and a cause–effect diagram is prepared for detailed analysis of machines and their failure. A performance assessment of equipment is also carried out based on the analysis. The key performance indicator used to measure performance is OEE since it measures the performance of equipment based on three different aspects i.e. availability, production rate and quality rate that contain all six main losses. After performance measures through OEE, a TPM plan is implemented initially in the critical section of the industry and then expanded on the whole industry.

8.3 ANALYSIS

8.3.1 QUESTIONNAIRE DATA ANALYSIS

The important questions from the questionnaire are discussed in this section. After data collection for TPM pillars, some findings are indicated. It is observed that no department of the industry has a robust maintenance plan. Only 39% of the respondent agreed of having a proper maintenance plan weekly or monthly. The improper maintenance planning leads to frequent breakdowns of machinery. The results are shown in Figure 8.2.

The next important question is about the involvement of workers in maintenance activities. This question is asked to know about the autonomous maintenance schedule of the industry. The responses observed only 28% of the worker involvement activity in maintenance activity and the most surprising that only 22% of workers get involved sometimes.

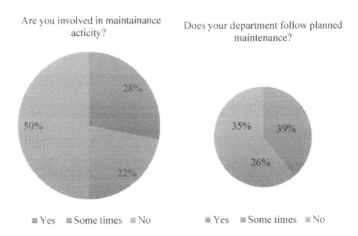

FIGURE 8.2   Maintenance planning and worker involvement.

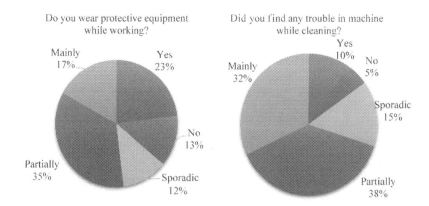

FIGURE 8.3   Safety and autonomous maintenance.

The third question is asked about the worker's capability and training to find the major and minor problem regarding the cleanliness of the machine. The response found that only 10% of workers were confident to locate a problem while cleaning. The fourth question is framed to know about worker safety and the usability of safety equipment because there exist major machines where the worker needs protective equipment. The responses found that only 23% of workers use protective equipment while working on the machine, as shown in Figure 8.3. Many questions are asked related to all major pillars of TPM, the distribution of question on TPM pillars are shown in Figure 8.4.

8.3.2   BREAKDOWN ANALYSIS OF EQUIPMENT

The analysis of machinery breakdown data is shown in Figure 8.5 and the results show that the machines from forging, broaching and shot blasting are critical and a strong maintenance plan is needed to avoid a frequent breakdown in this section.

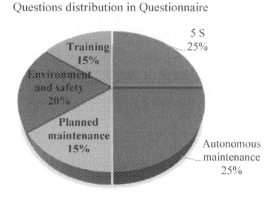

FIGURE 8.4   Analysis of TPM pillars.

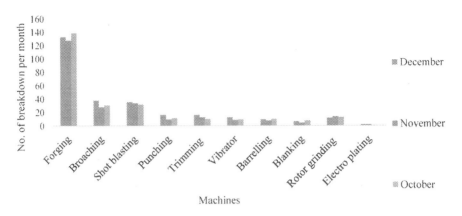

FIGURE 8.5   Breakdown/failure analysis.

These machines are further investigated to find out their cause of failure and corrective measures.

8.3.2.1   Analysis of the Cause of Machine Breakdown

Three critical machines having frequent failures are analyzed for a month. The machines forging hammer, broaching and shot blasting are having a maximum breakdown. Figure 8.6 shows the reasons for failure and the number of breakdowns per month for shot blasting machines. The cause–effect diagram is used for further investigation of the machine. The main failure in shot blasting machine is identified as the failure of the elevator, conveyor and rotors.

The main reason the stoppage at forging hammer is the die rib, metal sticking and high die setting time. The machine availability gets poorer due to frequent stoppages in forging hammer. The causes of failure in the forging hammer are shown in Figure 8.7. The high setting time has been reduced by implementing a SMED, a lean manufacturing tool.

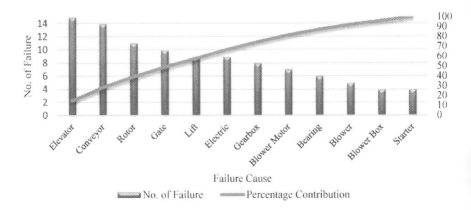

FIGURE 8.6   Failure cause of shot blasting machine.

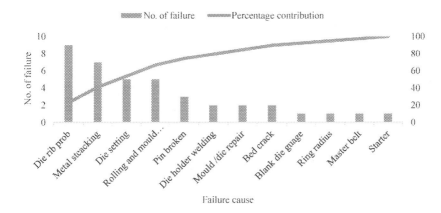

FIGURE 8.7 Failure cause of forging hammer.

In the case of the broaching machine, the failure of the control system and hydraulic pump, as well as leakage of coolant, are main causes (Figure 8.8).

8.3.3 MEASURING MACHINE PERFORMANCE

The performance of machines is calculated through well-known KPI i.e. OEE. Researches mention that the effectiveness of TPM is calculated by OEE to measure the performance of machines in the industry.

8.4 RESULTS AND DISCUSSION

8.4.1 OEE CALCULATION

OEE is the multiplication factor of these three factors denoted as availability (A), performance rate (P) and quality rate (Q). It integrates various important aspects of

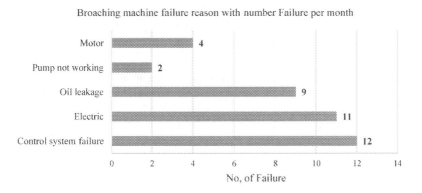

FIGURE 8.8 Causes of failure on broaching machine.

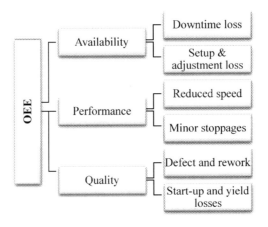

FIGURE 8.9   Losses and OEE.

the machine into a single performance-measuring tool. OEE also contains all the production losses in the system.

$$OEE = \text{Availability}(A) \times \text{Performance rate}(P) \times \text{Quality Rate}(Q)$$

The OEE in the current study is affected by six main losses of the industry.

The main losses that can occur in the industry are listed in Figure 8.9 and the data is collected for all the losses parameter.

$$\text{Availability }(A) = \frac{\text{Effective run time}}{\text{Planned production time}}$$

$$\text{Performance rate }(P) = \frac{\text{Idle Cycle Time} \times \text{Total Production}}{\text{Effective Run Time}}$$

$$\text{Quality Rate }(Q) = \frac{\text{Good production}}{\text{Total production}}$$

$$\text{Effective run time} = \text{Planned time} - \text{Breakdown time}$$

The OEE value of different machines is shown in Figure 8.10 for manually operated machines. The data is analyzed daily for one month and a graph has been plotted against OEE and machines.

The forging and trimming machines are having a maximum variation of OEE due to the high setting time of die and the maximum number of failures or minor stoppages of machines. For high die setting time, a SMED program is introduced to minimize the die set time to a single digit.

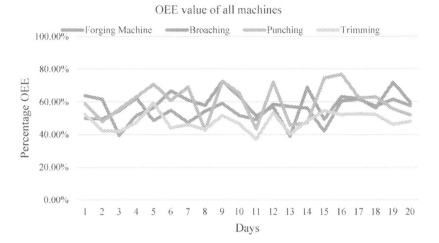

FIGURE 8.10 OEE for different machines.

8.4.2 SINGLE MINUTE EXCHANGE OF DIE

SMED is a lean tool used to minimize the die exchange time and in the current study, forging hammer takes 3–3.5 hours for die set which is reduced by 1 hour after implementation of SMED. The work distribution method is used for arranging tasks for minimizing die setting time. Some internal tasks are externalized and some unnecessary work and operator movement during die setting is eliminated (Figure 8.11). Die setting time is reduced from 193.2 to 125.6 minutes, which means 67.6 minutes of saving in die setting time. The SMED results also affect the availability of the forging hammer. This will indirectly improve the OEE of the forging

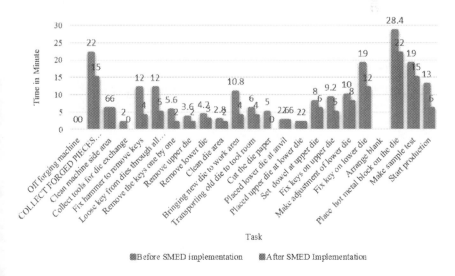

FIGURE 8.11 SMED result comparison.

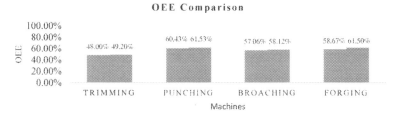

FIGURE 8.12   Performance comparison.

machine due to increased availability and production rate. By saving 67 minutes per setup, the machine can produce 978 more products per the setup of the die.

$$\text{Increase in production per setup} = \frac{67.6 \times 60}{4.12} = 984$$

Here, the cycle time for producing one forging piece is 4.12 second, and it is clear that the machine now has increased flexibility to produce more products in the same available time.

8.4.3 PERFORMANCE BEFORE AND AFTER TPM IMPLEMENTATION

After the successful implementation of TPM on select machines, the data is collected again and analyzed for an initial comparison (Figure 8.12). A reasonable improvement is observed; however, the performance will go on improving with time. The below graph shows the OEE improvement.

TABLE 8.1
Time Between Maintenance

Machine(s)	MTTF	MTTR	MTBF	Time Between Maintenance Needed (hours)
Hammer	64.48	0.48	64.00	22.98
Broaching	126.55	0.45	126.10	77.61
Shot blasting	69.09	0.45	68.64	21.13
Punching	321.23	0.45	320.78	197.45
Trimming	321.19	0.50	320.69	197.39
Vibrator	981.82	0.40	981.42	906.12
Barreling	626.40	0.33	626.07	578.03
Blanking	232.00	0.48	231.52	95.00
Rotor	214.15	0.47	213.69	87.69
Electroplating	2016.00	0.40	2015.60	310.16

8.4.4 PLANNED MAINTENANCE

Based on the current analysis, it is evident that the introduction of planned mainte-nance is highly mandatory, and a planned maintenance period is implemented using the reliability concept. This period will further extend on the performance of the machine goes improving.

Table 8.1 indicates that the forging and shot blasting machines need daily mainte-nance and given the criticality of the problem, immediate autonomous maintenance is proposed in these sections.

8.5 CONCLUSION

TPM is a powerful tool for the maintenance and care of machinery in any manu-facturing industry. This chapter discusses the implementation of TPM in a forging industry to maximize machinery availability, reducing non-value-added move-ments, and machinery breakdown time. After TPM implementation, OEE is used as a KPI for measuring the performance of a particular machinery section. The SMED approach is also implemented to reduce setup time in forging die setup. A significant savings of 67.6 minutes per setup is achieved that enables to produce 984 more products per day. Training seasons regarding follow autonomous maintenance were proposed to detect a fault on machinery. A significant improvement of OEE is achieved with this effort and most importantly, without any investment from the industry. These improvements are almost the same improvements achieved by Garg *et al.* (2016), Rodríguez-Méndez *et al.* (2015).

Thus, it is concluded that TPM help industry to achieve sustainable production goals by gaining economic and social benefit as a result of the improved effective-ness of machine and self-motive trained workers and fewer wastages. However, given time boundation and limited resources, only those improvements and tools were implemented that involved minimum time and there remains a significant scope of future work in this area.

REFERENCES

Adhiutama, A., Darmawan, R., & Fadhila, A. (2020). Total productive maintenance on the airbus part manufacturing. *Jurnal Bisnis dan Manajemen, 21*(1), 3–15.

Agustiady, T. K., & Cudney, E. A. (2018). Total productive maintenance. *Total Quality Management & Business Excellence,* 1–8, DOI:10.1080/14783363.2018.1438843.

Ahmad, N., Hossen, J., & Ali, S. M. (2018). Improvement of overall equipment efficiency of ring frame through total productive maintenance: A textile case. *The International Journal of Advanced Manufacturing Technology, 94*(1–4), 239–256.

Ahuja, I. P. S., & Khamba, J. S. (2009). Investigation of manufacturing performance achieve-ments through strategic total productive maintenance initiatives. *International Journal of Productivity and Quality Management, 4*(2), 129–152.

Borris, S. (2006). *Total productive maintenance.* USA: McGraw-Hill Companies, Inc.

Bhardwaj, A., Mor, R. S., & Nagar, J. (2018). Productivity gains through PDCA approach in an auto service station. *Proceedings of the 2nd IEOM European Conference on Industrial Engineering and Operations Management Paris, France,* July 26–27, 2018, pp. 2595–2602. ISBN: 978-1-5323-5945-3.

Brah, S. A., & Chong, W. K. (2004). Relationship between total productive maintenance and performance. *International Journal of Production Research, 42*(12), 2383–2401.

Chaudhuri, A., & Jayaram, J. (2019). A socio-technical view of performance impact of integrated quality and sustainability strategies. *International Journal of Production Research, 57*(5), 1478–1496.

Chen, P. K., Fortuny-Santos, J., Lujan, I., & Ruiz-de-Arbulo-López, P. (2019). Sustainable manufacturing: Exploring antecedents and influence of total productive maintenance and lean manufacturing. *Advances in Mechanical Engineering, 11*(11), 1-16.

Garg, G., Gupta, A., Mor, R. S., & Trehan, R. (2016). Execution of single minute exchange of die on corrugation machine in cardboard box manufacturing company: A case study. *International Journal of Lean Enterprise Research, 2*(2), 133–145, https://doi.org/10.1504/IJLER.2016.084580.

Gupta, P., & Vardhan, S. (2016). Optimizing OEE, productivity and production cost for improving sales volume in an automobile industry through TPM: A case study. *International Journal of Production Research, 54*(10), 2976–2988.

Gupta, K., Jain, N. K., & Laubscher, R. F. (2015). Spark erosion machining of miniature gears: A critical review. *The International Journal of Advanced Manufacturing Technology, 80*(9–12), 1863–1877.

Hami, N., Ibrahim, Y. M., Yamin, F. M., Shafie, S. M., & Abdulameer, S. S. (2019). The moderating role of sustainable maintenance on the relationship between sustainable manufacturing practices and social sustainability: A conceptual framework. *International Journal of Engineering and Advanced Technology, 8*(5C), 222–228.

Huang, S. H., Dismukes, J. P., Shi, J., Su, Q. I., Razzak, M. A., Bodhale, R., & Robinson, D. E. (2003). Manufacturing productivity improvement using effectiveness metrics and simulation analysis. *International Journal of Production Research, 41*(3), 513–527.

Ibrahim, Y. M., Hami, N., & Othman, S. N. (2019). Integrating sustainable maintenance into sustainable manufacturing practices and its relationship with sustainability performance: A conceptual framework. *International Journal of Energy Economics and Policy, 9*(4), 30.

Kigsirisin, S., Pussawiro, S., & Noohawm, O. (2016). Approach for total productive maintenance evaluation in water productivity: A case study at mahasawat water treatment plant. *Procedia Engineering, 154*, 260–267.

Kumar, S., Goyal, A., & Singhal, A. (2017). Manufacturing flexibility and its effect on system performance. *Jordan Journal of Mechanical & Industrial Engineering, 11*(2), 105–112.

Maalem, L. A. El, Ech-Chhibat, M. E. H., Zahiri, L., Ouajji, H., 2020. Total productive maintenance implementation in cables manufacturing: A case study. In *2020 1st Int. Conf. on Innovative Research in Applied Science, Engineering and Technology*, https://doi.org/10.1109/IRASET48871.2020.9092241

Mor, R. S., Bhardwaj, A., Singh, S., & Sachdeva, A. (2019). Productivity gains through standardization-of-work: Case of Indian manufacturing industry. *Journal of Manufacturing Technology Management, 30*(6), 899–919, https://doi.org/10.1108/JMTM-07-2017-0151.

Mor, R. S., Singh, S., Bhoria, R., & Bhardwaj, A. (2018). Role of value stream mapping in process development: A review. In *3rd North American International Conference on Industrial Engineering and Operations Management, Washington DC, USA, Sept. 27–29, 2018*, pp. 1938–1942. *ISBN: 978-1-5323-5946-0.*

Mwanza, B. G., & Mbohwa, C. (2015). Design of a total productive maintenance model for effective implementation: Case study of a chemical manufacturing company. *Procedia Manufacturing, 4*, 461–470.

Nakajima, S. (1988). *Introduction to TPM: Total productive maintenance.* (Translation). Productivity Press, Inc. (p. 129), ISBN-10: 0915299232, Shelton - CT 06484.

Rodríguez-Méndez, R., Sánchez-Partida, D., Martínez-Flores, J. L., & Arvizu-Barrón, E. (2015). A case study: SMED & JIT methodologies to develop continuous flow of stamped parts into AC disconnect assembly line in Schneider Electric Tlaxcala Plant. *IFAC-PapersOnLine*, *48*(3), 1399–1404.

Sahoo, S. (2020). Exploring the effectiveness of maintenance and quality management strategies in Indian manufacturing enterprises. *Benchmarking: An International Journal*, 27(4), 1399–1431.

Samadhiya, A., & Agrawal, R. (2020). Achieving sustainability through holistic maintenance-key for c. In *Proceedings of the International Conference on Industrial Engineering and Operations Management Dubai, UAE*, March 10–12, 2020.

Solke, N. S., & Singh, T. P. (2018). Analysis of relationship between manufacturing flexibility and lean manufacturing using structural equation modelling. *Global Journal of Flexible Systems Management*, *19*(2), 139–157.

Wudhikarn, R. (2012). Improving overall equipment cost loss adding cost of quality. *International Journal of Production Research*, *50*(12), 3434–3449.

9 A Multi-Criteria Decision-Making Model for Agricultural Machinery Selection

Ali Jahan, Alireza Panahandeh,
and Hadi Lal Ghorbani
Department of Industrial Engineering,
Semnan Branch, Islamic Azad University,
Semnan, Iran

CONTENTS

9.1 INTRODUCTION

The optimization process in field management is challenging due to the complexity of the existing agriculture machinery system and the variety of type, size, number and operating characteristics (Haffar & Khoury, 1992). Because combine harvesters are among the most expensive agricultural machines, all aspects and criteria should be taken into consideration when it comes to their selection. This issue becomes more important once we realize that 60% of the total costs of the agriculture sector spent on agricultural production is typically related to agricultural machinery and equipment (Dash, 2008). The growing complexity of engineering sciences and the new technologies has made it all the more necessary to use multi-criteria decision-making methods. The multi-criteria decision making (MCDM) means finding the most optimal solution out of all possible alternatives. In such problems, the high standards of judgment are inevitable. In order to make the right decision, a decision maker should evaluate several criteria and measure different options according to the criteria. MCDM models are further categorized into two types, namely multiple objective decision making (MODM) and multiple attribute decision making (MADM) (Jahan *et al.*, 2016). In general, multi-objective models are utilized in design while multi-criteria models are employed to select the best alternative. The main difference between MODM model and MCDM model is that the former is defined in the continuous decision-making space while the latter is defined in the discrete decision-making space. In multi-objective models, the number of options is infinite, but multi-attribute models have a number of options. The existence of conflicting criteria, each with a different significance and unit of measurement, is a characteristic of MCDM environment. In decision-making processes, decisions are further accompanied by uncertainties. For instance, in judging the quality of a product with multiple degrees of evaluation, the decision maker may not always be able to accurately evaluate the product to a certain degree; moreover, decisions may be ambiguous, and in certain cases, the initial data may also be inaccurate. In such cases, a multi-criteria decision-making method is required to model the inaccurate data (Beynon *et al.*, 2000).

Agricultural machinery is a growing technology that most farmers need in order to plant and harvest. Therefore, selecting the right machine is of utmost importance, and lack of definitive data in this field limits the use of conventional MCDM methods. In cases where it is difficult or impossible to obtain accurate data, interval data comprise a substantial portion of our retained information, increasing the accuracy of options prioritization. In addition, using interval numbers instead of fuzzy ones facilitates the decision-making process. Therefore, the purpose of this study was to propose an integrated model using interval data to evaluate and select the best harvesting mechanism.

The remainder of this paper is as follows: the second section contains literature review that provides an overview of research in the field of MCDM methods, the third section presents the research methodology, the fourth section describes the case study and discusses the results, and the fifth section concludes the whole paper with suggestions for future studies.

9.2 LITERATURE REVIEW

9.2.1 A BRIEF REVIEW OF ELECTRE METHOD

ELECTRE method is one of the most widely used methods for MCDM first introduced by Benayoun *et al.* (1966) and later renamed to ELECTRE I. As shown in Figure 9.1, this method has always been of interest to researchers. A major advantage of this method is that criteria information processing is almost compensatory.

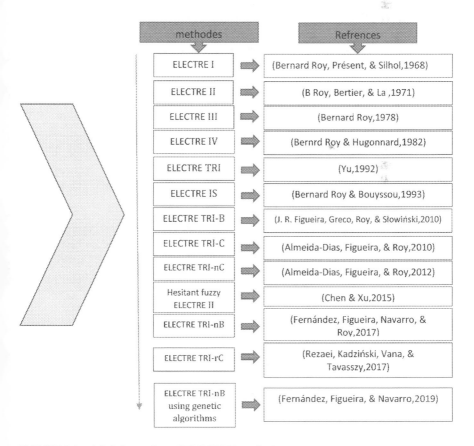

FIGURE 9.1   A brief overview of ELECTRE methods.

Thus, less important criteria of an option cannot be directly offset by other values of high importance criteria. On the other hand, the solution mechanism in ELECTRE method is not as extreme as purely non-compensatory methods.

Another type of ELECTRE, recently developed, is ELECTRE-IDAT method also used in the present research. Another upside of this method is the use of interval data to achieve an optimal response and focus on the target criteria (Jahan & Zavadskas, 2019).

9.2.2 MCDM in Agriculture Sector

Several papers have been published in the field of agriculture and tool selection based on MCDM methods. Amini and Asoodar (2016) selected the most suitable tractor using AHP method. Employing the exact same method, Rybacki (2011) addressed the topic of selecting a technical service station for agricultural tractors. Since many criteria and options are considered when selecting agricultural machinery, hierarchical analysis method can increase both the number of pairwise comparisons and the likelihood of inconsistency. Kohli et al. (2013) categorized research on the mechanism of cotton harvesting with different indices.

Using MULTIMOORA and WASPAS methods, Hafezalkotob et al. (2018) evaluated and selected agricultural machines for the purpose of economic growth and increased olive harvest efficiency. BEST-WORST method was further used to weight the existing criteria. Çakır (2018) employed both fuzzy simple multi-attribute rating technique (SMART) and fuzzy weighted axiomatic design (FWAD) to select the optimal tea dryer.

In another paper, Hu et al. (2019) proposed a dynamic maintenance network selection framework for agricultural machinery. Accordingly, Integrated Analytical Network Process (ANP) and Balanced Score Card (BSC) were opted for in order to select the maintenance and repair network.

9.2.3 Research Gap

Despite the extent of cotton fields, no report has been published on the ranking of cotton harvesters. Moreover, given the nature of the inaccurate data on the ranking of cotton harvesting mechanisms, it is necessary to use new MCDM methods. In this study, ELECTRE-IDAT and revised SIMOS were combined to select the most optimal alternative.

9.3 MATERIALS AND METHODS

Data collection from experts and their summarization in the form of a decision making matrix are highly important measures that can influence the results. Figure 9.2 summarizes the suggested methods for this study. The details are presented in five steps:

Step 1: Forming the decision making matrix
The decision-making matrix is formed based on the related studies conducted in this field and evaluation and selection of the best criteria by experts.

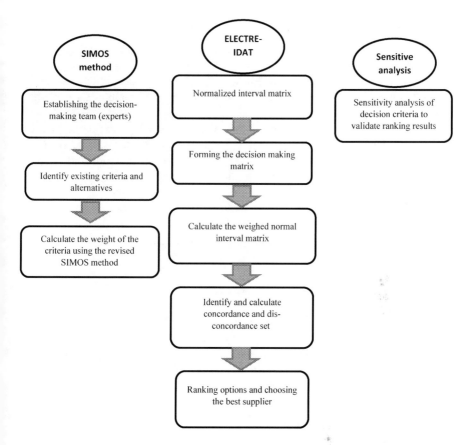

FIGURE 9.2 The proposed method for selecting the harvesting mechanism.

Step 2: Calculating the weights of the criteria using the revised SIMOS method

In this step, the criteria are weighted using expert opinions and revised SIMOS method.

The weighting methods are classified into objective and subjective groups; in the latter methods, the decision maker is responsible for criterion evaluation, while in the former, they are not. SIMOS method, a subgroup of mental methods, was used in the present study owing to its simplicity and efficiency (Siskos & Tsotsolas, 2015).

Step 3: Ranking options

In this section, the interval data is scaled and weighted using ELECTRE-IDAT method to rank the available options.

Step 4: Sensitivity analysis

After ranking the alternatives and selecting the most optimal one, the sensitivity analysis of the ranks is performed to change the importance of the criteria; the results are then evaluated and validated.

9.3.1 Revised SIMOS Method

SIMOS method is a weighting method that operates on a card playing process. Used for MCDM, this method is simple and understandable, especially for decision makers, and highly practical (Cavallaro, 2010), covering a wide range of decision problems (Shanian et al., 2008). The steps of the revised SIMOS method are explained below (Figueira & Roy, 2002).

The decision maker receives a number of cards on which the name of each criterion is written. There are other cards called white cards. The decision maker ranks the cards from the least important to the most important. If some criteria are of equal importance to the decision maker, they fall into one column or below one another; if the importance of a criterion is twice the other, a white card is placed between them. Likewise, for the importance of three-fold, two white cards are used. In REVISED SIMOS method, the information obtained from the decision maker is used to calculate the weight of the criteria as follows:

Step 1: Defining the number of scale units

$$S_r = S_r' + 1 \ r = 1, 2, 3, \ldots, \bar{n} - 1 \tag{1}$$

According to the formula above, if there is no white card between the two criteria, S_r interval between them is one unit. Also, if a white card falls between the criteria, S_r doubles, and so on.

Step 2: Calculating the sum of the units between the first and last criterion shown by S.

$$S = \sum_{r=1}^{n} S_r \tag{2}$$

Step 3: In this step, the decision maker is asked to determine whether the criterion with the first priority is several times the value of the last criterion. The answer to this question specifies the Z value, which is used to determine the constant interval between the criteria weights or the subset of criteria determining the u value of this interval.

$$u = \frac{Z-1}{S} \tag{3}$$

Step 4: Calculating the abnormal weight:

$$P(r) = 1 + u\left(S_0 + \ldots + S_{r-1}\right) \ r = 1, \ 2, \ \ldots, \ \bar{n} \ S_0 = 0 \ P(r) = 1 \tag{4}$$

Step 5: Calculating the normal weight

For each available criterion, non-normalized weight is shown with P_j', which is the same weight in the criteria pertaining to the same subdivided columns or cards. In other words, for each criterion, $P'_j = P(r)$. The normalized weight is represented with P_j^* calculated by formula (5).

$$P_j^* = \frac{P'_j}{P'} \tag{5}$$

$$P' = \sum_{j=1}^{n} P'_j$$

9.3.2 ELECTRE METHOD FOR INTERVAL DATA AND TARGET CRITERIA (ELECTRE-IDAT)

In classical ELECTRE methods, the available options are ranked by the crisp data; however, it is sometimes difficult to accurately determine the actual values of the options against the criteria; therefore, we used the interval data in the present research. ELECTRE-IDAT method is a systematic method for interval data and target-based criteria (Jahan & Zavadskas, 2019). This method is capable of dealing with variety of criteria. The four types of criteria are outlined as follow. *Target criteria:* These types of criteria are focused on a specific purpose or measure, and their purpose is to achieve that attribute or size. For instance, if a target-type criterion indicates a value of 30, the option closest to the given number gains more weight. *Bounded criteria* have upper or lower bounds. In the *cost metrics*, the lowest value option is selected; for example, the purchase price factor is the cost type. The *profit criteria* are the opposite of the cost ones. In these criteria, the options with the highest value are accepted. For instance, a criterion of income is a criterion of profit.

In accordance with the above definitions and explanations, the steps of ELECTRE-IDAT method are as follows:

Step 1: Forming the decision matrix on MCDM problem using the interval data shown in Table 9.1.

TABLE 9.1
MADM Problem with Interval Data

Weights	W_1	W_2	...	W_n
Criteria	C_1	C_2	...	C_n
A_1	$\left[x_{11}^L, x_{11}^U\right]$	$\left[x_{12}^L, x_{12}^U\right]$	...	$\left[x_{1n}^L, x_{1n}^U\right]$
A_2	$\left[x_{21}^L, x_{21}^U\right]$	$\left[x_{22}^L, x_{22}^U\right]$	...	$\left[x_{2n}^L, x_{2n}^U\right]$
A_3	$\left[x_{31}^L, x_{31}^U\right]$	$\left[x_{32}^L, x_{32}^U\right]$	...	$\left[x_{3n}^L, x_{3n}^U\right]$
$\vdots$	$\vdots$	$\vdots$		$\vdots$
A_m	$\left[x_{m1}^L, x_{m1}^U\right]$	$\left[x_{m2}^L, x_{m2}^U\right]$	...	$\left[x_{mn}^L, x_{mn}^U\right]$

Step 2: In this step, the decision matrix is normalized using Equations (6) and (7).

$$n_{ij}^L = \text{Min}\left[1 - \frac{\left|x_{ij}^L - T_j\right|}{\max\left\{\left|B - T_j\right|, \left|A - T_j\right|\right\}}, 1 - \frac{\left|x_{ij}^U - T_j\right|}{\max\left\{\left|B - T_j\right|, \left|A - T_j\right|\right\}}\right] \tag{6}$$

$$n_{ij}^U = \text{Max}\left[1 - \frac{\left|x_{ij}^L - T_j\right|}{\max\left\{\left|B - T_j\right|, \left|A - T_j\right|\right\}}, 1 - \frac{\left|x_{ij}^U - T_j\right|}{\max\left\{\left|B - T_j\right|, \left|A - T_j\right|\right\}}\right] \tag{7}$$

The above equations are used for the interval data, where $i = 1, 2, ..., m$ and $j = 1, 2, ..., n$, and [A, B] are the data ranges belonging to each criterion in the available option. T_j indicates the optimal position of the data in each criterion. According to the definition of T_j as stated below, if the criterion is a cost type, T_j is considered as the lower value (A). In profit criteria, T_j is the maximum value (B) and if in terms of target criteria, it is the same target value as T_j. The following are the brief steps for T_j selection:

A	Less is better (cost criteria)
$T_j = B$	More is better (benefit criteria)
Between A and B	Approaching the specific value is desirable (target criteria)

Step 3: Determining the Weighted Normal Interval Matrix
In this step, using Equations (8) and (9), the weights calculated by the revised SIMOS method are multiplied by the normalized matrix, and the normalized weighted matrix is then calculated. Therefore, we name $\bar{V}_{ij} = \left[V_{ij}^L, V_{ij}^U\right]$ the 'weighted normal matrix'.

$$V_{ij}^L = w_j n_{ij}^L \quad i = 1, 2, ..., m \quad j = 1, 2, ..., n \tag{8}$$

$$V_{ij}^U = w_j n_{ij}^U \quad i = 1, 2, ..., m \quad j = 1, 2, ..., n \tag{9}$$

Step 4: Identifying and selecting the set of concordance and dis-concordance criteria in the interval data
According to matrix $\bar{V}_{ij}$, the concordance dataset $(S_{k,1})$ and the dis-concordance dataset $(D_{k,1})$ are identified, where $k, 1 = 1,2,..., m; 1 \neq k. S_{k,1}$ is selected using the following equations and based on Table 9.2.

$$S_{k,l} = \left\{j \mid \bar{V}_{kj} \geq \bar{V}_{lj}\right\} = \left\{j \mid \left[V_{kj}^L, V_{kj}^U\right] \geq \left[V_{lj}^L, V_{lj}^U\right]\right\} \tag{10}$$

$$P\left(\bar{V}_{kj} \geq \bar{V}_{lj}\right) = P\left(\left[V_{kj}^L, V_{kj}^U\right] \geq \left[V_{lj}^L, V_{lj}^U\right]\right)$$
$$= \frac{\text{Max}\left\{0, V_{kj}^U - V_{lj}^L\right\} - \text{Max}\left\{0, V_{kj}^L - V_{lj}^U\right\}}{V_{kj}^U - V_{kj}^L + V_{lj}^U - V_{lj}^L} \tag{11}$$

TABLE 9.2
Guidelines for Determining Concordance Interval Sets

If	Then
$if\left(V_{kj}^{L},V_{kj}^{U} \text{ and } V_{lj}^{L},V_{lj}^{U}\right) \rightarrow P\left(\bar{V}_{kj} \geq \bar{V}_{lj}\right)=1$	j is added to $S_{k.l}$
$if\left(V_{kj}^{L} \geq V_{lj}^{U}\right) \rightarrow P\left(\bar{V}_{kj} \geq \bar{V}_{lj}\right)=1$	j is added to $S_{k.l}$
$if\left(P\left(\bar{V}_{kj} \geq \bar{V}_{lj}\right) \geq 0.5\right)$	j is added to $S_{k.l}$

The following criteria are not included in the set of concordance ($S_{k,l}$) in the interval dis-concordance dataset ($D_{k,l}$).

$$D_{k,l} = \left\{ j \mid \left[V_{kj}^{L},V_{kj}^{U} \right] < \left[V_{lj}^{L},V_{lj}^{U} \right] \right\} = J - S_{k,l} \qquad (12)$$

Step 5: Calculating the concordance interval Matrix Values
The concordance interval matrix is shown with *CI* calculated and displayed as follows:

$$CI = \begin{vmatrix} - & CI_{1,2} & CI_{1,3} & \dots & \dots & CI_{1,m} \\ CI_{2,1} & - & . & \dots & \dots & CI_{2,m} \\ CI_{3,1} & . & - & \dots & \dots & CI_{3,m} \\ . & . & . & \dots & \dots & . \\ . & . & . & \dots & \dots & CI_{m-1,m} \\ . & . & . & \dots & - & . \\ CI_{m,1} & CI_{m,2} & CI_{m,3} & \dots & CI_{m,m-1} & - \end{vmatrix}$$

where (13)

$$CI_{k,l} = \sum_{j \in S_{k,l}} W_j$$

Step 6: Calculating the concordance interval matrix defined as follows:

$$DI = \begin{vmatrix} - & DI_{1,2} & DI_{1,3} & \dots & \dots & DI_{1,m} \\ DI_{2,1} & - & . & \dots & \dots & DI_{2,m} \\ DI_{3,1} & . & - & \dots & \dots & DI_{3,m} \\ . & . & . & \dots & \dots & . \\ . & . & . & \dots & \dots & CI_{m-1,m} \\ . & . & . & \dots & - & . \\ DI_{m,1} & DI_{m,2} & DI_{m,3} & \dots & DI_{m,m-1} & - \end{vmatrix}$$

where (14)

$$DI_{k,l} = \frac{\underset{j \in D_{k,l}}{\text{Max}} \left[V_{kj}^L, V_{kj}^U \right] - \left[V_{lj}^L, V_{lj}^U \right]}{\underset{j \in J}{\text{Max}} \left[V_{kj}^L, V_{kj}^U \right] - \left[V_{lj}^L, V_{lj}^U \right]}$$

Step 7: Calculating the net values of concordance (C_i) and dis-concordance (D_i) for the available options:
Pure concordance index (C_i)

$$C_i = \sum_{k=1}^m CI_{i,k} - \sum_{k=1}^m CI_{k,i} \; (i \neq k) \tag{15}$$

Pure dis-concordance index (D_i)

$$D_i = \sum_{k=1}^m DI_{i,k} - \sum_{k=1}^m DI_{k,i} \; (i \neq k) \tag{16}$$

Step 8: Calculating E_i index values and ranking options
In this step, after estimating concordance and dis-concordance indices, E_i index is calculated using Equation (17). Given this equation we have:

$$C^- = \text{Min } C_i, C^+ = \text{Max } C_i, D^- = \text{Min } D_i, D^+ = \text{Max } D_i$$

$$E_i = \begin{cases} \left[\dfrac{D_i - D^-}{D^+ - D^-} \right] & \textit{if } C^+ = C^- \\[3mm] \left[\dfrac{C_i - C^-}{C^+ - C^-} \right] & \textit{if } D^+ = D^- \\[3mm] \left[\dfrac{C_i - C^-}{C^+ - C^-} \right] \gamma + \left[\dfrac{D^+ - D_i}{D^+ - D^-} \right] (1 - \gamma) & \text{otherwise} \end{cases} \tag{17}$$

9.4 CASE STUDY: SELECTION OF COTTON HARVESTING MECHANISM

This section outlines a case study in the field of agriculture with the aim of selecting the most efficient mechanism for harvesting cotton via the proposed method. To select the most optimal cotton harvesting mechanism, the present study investigated the three existing criteria, namely harvest efficiency, trash bin content and Gin turn out. Cotton harvesting is comprised of spindle, brush and paddle, and Finger, explained in more detail below.

9.4.1 COTTON COLLECTORS

Cotton collectors only separate fibers from the pod, and the machine is of a mounted type with one or more rotating cylindrical striping, on which several needles are inserted (zigzag); when the cylinders are in rotation, the fingers are bent into the pod, pulling the cotton out of it; next, the fingers pass through the vertical cylindrical plates, called the cotton holder, and the cotton is taken and guided to the basket; after that, the fingers rotate to another cylinder whose plates are always wet, so the residue of sticking cotton is also wiped. In addition to its high costs, cotton collector crushes the field soil because it has to enter the field several times.

9.4.2 POD COLLECTORS

In pod collectors, the pods are separated from the cotton by the machine of the finger or roller brush and paddle type. The benefit of this mechanism is that the machine is no longer required to enter the field several times, and the downside is the long time required to convert pod to cotton. Figure 9.3 shows the types of cotton and pod collectors. If the area is windy, pod collector is used while if it is not, cotton collector is utilized.

9.4.3 THE STUDIED CRITERIA FOR SELECTION OF HARVESTING MECHANISM

9.4.3.1 Picking Efficiency

Generally, different types of picking efficiency have been reported; for instance, for a spindle collector it is between 95% and 98%, but there are some manufacturers

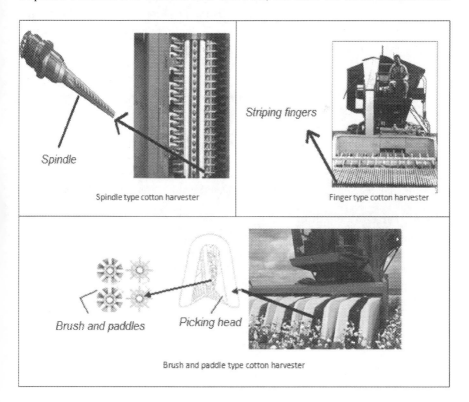

FIGURE 9.3 Different mechanisms of cotton harvesting.

with 20% waste. Several problems can lead to picking efficiency loss such as not planting in a row or picking too soon after using defoliants. This index has a positive aspect and is more desired (Kohli *et al.*, 2013).

9.4.3.2   Trash Bin Container

The content of a trash bin can be measured by various laboratory instruments and high-volume measurement systems. Complex devices have been developed using image analysis and color differentiation to more accurately measure the level and type of trash. The content of trash bin includes external materials such as wood, thorns, leaves, shells and so forth. This index has a negative aspect and is less desired.

9.4.3.3   Gin Turn Out

Gin turn out is the percentage of cotton obtained following the Gin turn out process. The percentage of Gin turn out depends on the yield of the product and the content of the trash bin. This index has a positive side and most of it is desired.

9.4.4   AVAILABLE OPTIONS TO CHECK PRIORITIZATION FOR SELECTION

9.4.4.1   Spindle Mechanism

Spindle mechanism, known as 'spinning wheel', is a cotton harvester. The spindle has spindle rows that remove cotton from the plant by rotating at high speeds. The cotton seed is removed from the spindles by anti-rotation and then poured into the tank. The spindle-type machine was designed to remove the product using a spindle without injuring the foliage of the plant (Muthamilselvan *et al.*, 2007).

9.4.4.2   Brush and Paddle Mechanism

Brush cleaner consists of two cylinders and rows of bats including a hard rubber brush that rotates and takes the cotton stem. The stems further pass through the cylinders, and the entire pod, whether open or unopened, is removed from the stem. As far as repairing and maintaining are concerned, the brush type mechanism is less expensive compared with the spindle type. The brush type picker removes more materials than the spindle type, but the expected yield is less (Spurlock *et al.*, 1991).

9.4.4.3   Finger Mechanism

Finger mechanism removes the cotton pod from the plant by moving upwards and exerting an upward force while the plant is positioned between the fingers of the harvester. Colwick *et al.* (1979) reported that compared with spindle cleaner, the finger-type was more effective; however, the quality of the product was one degree lower in finger cleaner.

9.5   RESULTS AND DISCUSSION

9.5.1   IDENTIFYING THE CRITERIA AND OPTIONS

Tables 9.3 and 9.4 present the options and criteria for cotton harvesting mechanism based on library studies conducted in this field and the material presented in Section 4.

TABLE 9.3
Criteria for the Selection of Harvesting Mechanism

Number	Criteria for Cotton Harvester	Nature of the Data	Symbol
1	Picking efficiency	Quantitative and interval	c_1
2	Trash content	Quantitative and interval	c_2
3	Gin turnout	Quantitative and interval	c_3

9.5.2 CALCULATING THE WEIGHT OF THE CRITERIA USING THE REVISED SIMOS METHOD

In this step, the experts are provided with the criteria cards and white cards, shown in Figure 9.4. Then, by assuming $z = 10$, the value of P_r is calculated (Table 9.5). Finally, using Equation (5), we calculate the normalized weights of the existing criteria according to Table 9.6.

Therefore, each existing criterion is calculated using the revised SIMOS method. According to the above table, we have:

$c_1 = 0.067, c_2 = 0.267, c_3 = 0.667$

9.5.3 RANKING THE OPTIONS USING ELECTRE-IDAT METHOD

Step 1: Forming the decision-making matrix
By collecting data from previous studies, we formed the decision-making matrix (Table 9.7). This matrix identifies the properties and types of each criterion and uses interval data to evaluate each alternative.
Step 2: Normalized interval matrix formation:
In this step, the data are scaled (Table 9.8) according to Equations (6) and (7).
Step 3: Forming a weighted normalized interval matrix
Using Equations (8) and (9), weighted normal matrix is calculated as shown in Table 9.9.
Step 4: Identifying and determining the concordance and dis-concordance criteria set
In this step, using the framework of Table 9.2, the set of concordance and dis-concordance matrices criteria is identified in relation to both options. The results are shown in Table 9.10.

TABLE 9.4
Harvesting Mechanism for Cotton

Number	The title alternative for cotton harvester	Alternative symbol
1	Brush and paddle harvesting mechanism	M_1
2	Spindle harvesting mechanism	M_2
3	Finger harvesting mechanism	M_3

c1		c2				c3

FIGURE 9.4 Schematic of card played by experts.

TABLE 9.5
Calculation Process for Weighting

r	1	2	3	S
s'_r	1	3	-	
s_r	2	4	-	6
p_r	1	4	10	-

TABLE 9.6
Normalized Weightings

Criteria	c_1	c_2	c_3	p'
J	1	2	3	
p_r	p_1	p_2	p_3	
p'_r	1	4	10	15
p^*_r	0/067	0/267	0/667	1/000

TABLE 9.7
The Decision Matrix

Type of Criteria	Benefit		Cost		Benefit	
Criteria	c_1		c_2		c_3	
Nature of data	Quantitative and interval		Quantitative and interval		Quantitative and interval	
Weight	0/067	0/067	0/267	0/267	0/667	0/667
M_1	90	100	15	20	30	35
M_2	75	80	0	10	35	40
M_3	95	100	25	100	30	35
A	75		0		30	
B	100		100		40	
T	100		0		40	

TABLE 9.8
Normalized Decision Matrix

	C_1		C_2		C_3	
M_1	0/6	1	0/8	0/85	0	0/5
M_2	0	0/2	0/9	1	0/5	1
M_3	0/8	1	0	0/75	0	0/5

TABLE 9.9

Weighted Normalized Interval Matrix

	C_1		C_2		C_3	
M_1	0/040	0/067	0/213	0/227	0/000	0/333
M_2	0/000	0/013	0/240	0/267	0/333	0/667
M_3	0/053	0/067	0/000	0/200	0/000	0/333

Step 5: Calculating the concordance and dis-concordance matrices
In this step, the concordance and dis-concordance matrix indices are calculated. In this regard, Equations (10)–(12) were used to calculate the ratio and
form the matrices shown in Tables 9.11 and 9.12.

Step 6: Final ranking of options
After forming the concordance and dis-concordance matrices, the values of C_i and D_i were calculated through Equations (15) and (16).
Next, using the obtained data, the values of E_i index we calculated
based on Equation (17). It is to be noted that the options were ranked
by considering E_i values in descending order (larger to smaller data).
Table 9.13 shows the calculated values, including the final ranking of the
options.

According to the results, we observed that $M_2 > M_1 > M_3$, hence the fact
that M_2 was selected as the most optimal alternative. According to the
above results and the studied options in the case study, we have:

> Finger harvesting mechanism > Brush and paddle harvesting
> mechanism > Spindle harvesting mechanism

9.5.4 SENSITIVITY ANALYSIS

Sensitivity analysis is an important tool for measuring the validity of results in
research. Sensitivity analysis is a method for observing the effect of changes in
some model parameters on other elements. This is generally done by replacing the
weight of each criterion (Yazdani *et al.*, 2017). Therefore, as shown in Table 9.14, the

TABLE 9.10

The Set of the Concordance and Dis-Concordance Criteria

	Concordance Criteria Set	Dis-Concordance Criteria Set
M_1 vs. M_2	C_1	C_2, C_3
M_1 vs. M_3	C_2, C_3	C_1
M_2 vs. M_3	C_2, C_3	C_1

TABLE 9.11

Concordance Matrix

	M_1	M_2	M_3
M_1	-	0/067	0/933
M_2	0/933	-	0/933
M_3	0/067	0/067	-

TABLE 9.12

Dis-Concordance Matrix

	M_1	M_2	M_3
M_1		1/000	0/056
M_2	0/140		0/160
M_3	1/000	1/000	

TABLE 9.13

The Final Ranking of Options

	C_i	d_i	$\left[\dfrac{C_i - C^-}{C^+ - C^-}\right]$	$\left[\dfrac{D_i - D^-}{D^+ - D^-}\right]$	E_i	Ranking
M_1	0/000	-0/084	0/5	0/463647959	0/73182398	2
M_2	1/733	-1/700	1	0	1	1
M_3	-1/733	1/784	0	1	0/5	3

TABLE 9.14

Calculated E_i options for $Z = [1, 10]$

Z	M_1	M_2	M_3
1	0/5000	1/5000	0/4784
2	0/5000	1/0106	0/5000
3	0/6429	1/0000	0/5000
4	0/6846	1/0000	0/5000
5	0/7036	1/0000	0/5000
6	0/7143	1/0000	0/5000
7	0/7211	1/0000	0/5000
8	0/7258	1/0000	0/5000
9	0/7292	1/0000	0/5000
10	0/7318	1/0000	0/5000

TABLE 9.15
BORDA Results

$Z = [1-10]$	M_1	M_2	M_3	Dominance	Borda	Ranking
M_1	-	0	1	1	2	2
M_2	1	-	1	2	4	1
M_3	0	0	-	0	0	3
ΣR	1	0	2			

Z values of the revised SIMOS method ranged from 1 to 10 and the obtained weights were used according to the research methodology by ELECTRE-IDAT method; E_i values of each alternative were further calculated. Next, using Borda method, shown in Table 9.15, the dominance values of each option were determined and ultimately ranked. According to the results, no significant change was observed in the ranking of options, corroborating the stability of the model.

9.6 CONCLUSION AND RECOMMENDATION FOR FURTHER STUDY

The objective was to select a supplier in the field of supply chain management; therefore, we evaluated and selected the best supplier using MCDM methods. Selecting the most appropriate harvesting mechanism requires considering several criteria. In this study, three indices were selected, namely harvest efficiency, trash bin content and Gin turn out; three harvesting mechanisms were further compared. Due to the lack of complete information in the field of cotton harvesting mechanism, we encountered uncertainties, hence the use of interval data as one possible solution. Therefore, revised SIMOS for weighting criteria and ELECTRE-IDAT were combined in order to rank and select the best option. ELECTRE-IDAT method allowed us to easily work on a variety of criteria such as profit, cost, goal and bound so as to make more accurate and efficient decisions. The study results showed that the spindle mechanism was the most optimal mechanism for harvesting cotton; the next priorities were brush, paddle and Finger mechanisms.

According to the study results, here are some suggestions for future studies:

• Using this methodology in fuzzy environments and comparing it with interval and deterministic data
• Analyzing ELECTRE-IDAT method in different case studies and comparing and analyzing its results
• Using other MCDM methods and evaluating and comparing them with the present methodology
• Considering crop indices such as spacers, plant height, and plant population.

REFERENCES

Almeida-Dias, J., Figueira, J. R., & Roy, B. (2010). Electre Tri-C: A multiple criteria sorting method based on characteristic reference actions. *European Journal of Operational Research*, *204*(3), 565–580, Elsevier.

Almeida-Dias, J., Figueira, J. R., & Roy, B. (2012). A multiple criteria sorting method where each category is characterized by several reference actions: The Electre Tri-nC method. *European Journal of Operational Research*, *217*(3), 567–579. Elsevier.

Amini, S., &Asoodar, M. A. (2016). Selecting the most appropriate tractor using analytic hierarchy process – An Iranian case study. *Information Processing in Agriculture*, *3*(4), 223–234. Elsevier.

Benayoun, R., Roy, B., & Sussman, B. (1966). ELECTRE: Une méthode pour guider le choix en présence de points de vue multiples, SEMA (Metra international), direction scientifique, note de travail n 49', Paris, France.

Beynon, M., Curry, B., & Morgan, P. (2000). The Dempster–Shafer theory of evidence: An alternative approach to multicriteria decision modelling. *Omega*, *28*(1), 37–50. Elsevier.

Çakır, S. (2018). An integrated approach to machine selection problem using fuzzy SMART-fuzzy weighted axiomatic design. *Journal of Intelligent Manufacturing*, *29*(7), 1433–1445. Springer.

Cavallaro, F. (2010). A comparative assessment of thin-film photovoltaic production processes using the ELECTRE III method. *Energy Policy*, *38*(1), 463–474. Elsevier.

Chen, N., &Xu, Z. (2015). Hesitant fuzzy ELECTRE II approach: A new way to handle multicriteria decision making problems. *Information Sciences*, *292*, 175–197. Elsevier.

Colwick, R. F., Rayburn, S., & Barker, G. L. (1979). Row spacing, short season and stripper harvesting of cotton in northeast Mississippi (Vol. 94): AR, SEA, USDA

Dash, R. C. (2008). A computer model to select optimum size of farm power and machinery for paddy-wheat crop rotation in northern India. Agricultural Engineering International, 1-12. CIGR Journal.

Fernández, E. *et al.* (2017). ELECTRE TRI-nB: A new multiple criteria ordinal classification method. *European Journal of Operational Research*, *263*(1), 214–224. doi: 10.1016/J.EJOR.2017.04.048.

Fernández, E., Figueira, J. R., & Navarro, J. (2019). An indirect elicitation method for the parameters of the ELECTRE TRI-nB model using genetic algorithms. *Applied Soft Computing*, *77*, 723–733. Elsevier. doi: 10.1016/J.ASOC.2019.01.050.

Figueira, J. R. *et al.* (2010). ELECTRE methods: Main features and recent developments. In *Handbook of Multicriteria Analysis*. Springer, pp. 51–89.

Figueira, J., & Roy, B. (2002). Determining the weights of criteria in the ELECTRE type methods with a revised Simos' procedure. *European Journal of Operational Research*. 139(2), 317-326. doi: 10.1016/S0377-2217(01)00370-8.

Hafezalkotob, A. *et al.* (2018). A decision support system for agricultural machines and equipment selection: A case study on olive harvester machines. *Computers and Electronics in Agriculture*, *148*, 207–216. Elsevier.

Haffar, I., & Khoury, R. (1992). A computer model for field machinery selection under multiple cropping. *Computers and Electronics in Agriculture*, *7*(3), 219–229. Elsevier.

Hu, Y. *et al.* (2019). An ANP-multi-criteria-based methodology to construct maintenance networks for agricultural machinery cluster in a balanced scorecard context. *Computers and Electronics in Agriculture*, *158*, 1–10. Elsevier. doi: 10.1016/J.COMPAG.2019.01.031.

Jahan, A., Edwards, K. L., & Bahraminasab, M. (2016). *Multi-criteria decision analysis for supporting the selection of engineering materials in product design.* Butterworth-Heinemann.

Jahan, A., & Zavadskas, E. K. (2019). ELECTRE-IDAT for design decision-making problems with interval data and target-based criteria. *Soft Computing*, *23*(1), 129–143. Springer.

Jones, J. W. *et al.* (1979). Optimum time for harvesting cotton: A new concept. *Transactions of the ASAE. American Society of Agricultural and Biological Engineers. 22*(2), 291–296.

Kohli, S. S. *et al.* (2013). Multiple attribute decision making for selection of mechanical cotton harvester. *Scientific Research and Essays. Academic Journals. 8*(48), 2318–2331.

Muthamilselvan, M. *et al.* (2007). Mechanical picking of cotton – A review. *Agricultural Reviews. Agricultural Research Communication Centre, 28*(2), 118–126.

Rezaei, J. *et al.* (2017). Embedding carbon impact assessment in multi-criteria supplier segmentation using ELECTRE TRI-rC. *Annals of Operations Research*, 1–23. Springer. https://doi.org/10.1007/s10479-017-2454-y

Roy, B. (1978). ELECTRE III: Un algorithme de classements fondé sur une représentation floue des préférences en présence de critères multiples. *Cahiers du CERO, 20*(1), 3–24.

Roy, B., Bertier, P., & La méthode ELECTRE, I. I. (1971). Note de travail 142. *SEMA-METRA, Metra-International.*

Roy, B., & Bouyssou, D. (1993). *Aide multicritère à la décision: Méthodes et cas.* Economica Paris.

Roy, B., &Hugonnard, J.-C. (1982). Ranking of suburban line extension projects on the Paris metro system by a multicriteria method. *Transportation Research Part A: General, 16*(4), 301–312. Elsevier.

Roy, B., Présent, M., & Silhol, D. (1986). A programming method for determining which Paris metro stations should be renovated. *European Journal of Operational Research, 24*(2), 318–334. Elsevier.

Rybacki, P. (2011). Investigation of the decision-making process of selection of service station for agricultural tractors using the AHP method. *Journal of Research and Applications in Agricultural Engineering. 56*(2), 126–130.

Shanian, A. *et al.* (2008). A new application of ELECTRE III and revised Simos' procedure for group material selection under weighting uncertainty. *Knowledge-Based Systems, 21*(7), 709–720. Elsevier.

Siskos, E., & Tsotsolas, N. (2015). Elicitation of criteria importance weights through the Simos method: A robustness concern. *European Journal of Operational Research, 246*(2), 543–553. Elsevier.

Spurlock, S. R., Parvin Jr, D. W., & Cooke Jr, F. T. (1991). *Impacts of scrapping on harvest costs. In Proceedings-Beltwide Cotton Conferences (USA).*

Wójcik-Leń, J. *et al.* (2019). Studies regarding correct selection of statistical methods for the needs of increasing the efficiency of identification of land for consolidation— A case study in Poland. *Land Use Policy, 87,* 104064. Pergamon. doi: 10.1016/J.LANDUSEPOL.2019.104064.

Yazdani, M. *et al.* (2017). A group decision making support system in logistics and supply chain management. *Expert Systems with Applications, 88,* 376–392.

Yu, W. (1992). 'ELECTRE TRI : Aspects méthodologiques et manuel d'utilisation', Dauphine: *Université Paris.*

Section 4

Circular Economy and
Related Area Practices in
Operations Management

10 Challenges in Implementing Green Supply Chain Management in SMEs: A Case Study of a South Korean Company

Arvind Upadhyay
Brighton Business School, University
of Brighton, Brighton UK

Jeagyung Seong
Gudeul Technology, Wonju South Korea

CONTENTS

10.1 INTRODUCTION

Environmental management is a critical strategic area to maximise a firm's performance, and its importance is increasing steadily (Zhu & Sarkis, 2006). As a result of the implementation of Environmental Management Systems (EMS) for businesses, an increase in the use of EMS tools has been noted. Apart from internal environmental improvement, it is often the case that organisations are employing EMS guidelines past the boundaries of their factories and these can be seen in their supply chain networks (Sarkis, 2003). Effects on the environment can be seen throughout a number of points in a product's life cycle, including material sourcing and selection, manufacturing processes, reuse, product delivery and disposal (Ojo et al., 2012). Green supply chain management (GSCM), together with other connected standards, has turned into a crucial area for firms to create new profit and market advantages through reduction in environmental hazards and increased efficiency (Sarkis, 1995).

Nevertheless, there are certain parties that view this topic differently, as it is in contrast with the idea that environmental and social integration are mutually exclusive to financial gain. This disparity is particularly notable when comparing the views of business academics and practice. It is suggested that there is an inherent clash between protecting the environment and economic prosperity. The idea is based on the notion that the increasing need for organisations to protect the environment will cause a rise in their expenses, thus providing fewer resources to boost productivity and causing a drop in market competitiveness (Palmer et al., 1995). A McKinsey & Company (2008) reported that the vast majority (4 out of every 5) of 2192 executives contacted believed that there would be an environmental problem of some kind (most likely climate change-related, in their view) in the next five years in the regions where their firms work. Environmentally proactive firms attempt to structure these guidelines and even encourage stricter rulings, which could cause difficulty for their less environmentally conscious rivals. Moreover, 81% of the executives contacted state that businesses must be active in combating climate change, as this is seen as a societal trend with the highest chance of impacting shareholder value in the next 5 years, together with other environmental and social concerns. Lee, K. H. (2009) stated an important question for executives with regards to corporate EMS or GSCM is how to implement environmental decision making into their business practices, with profitable outcomes. However, it should be noted that business and management research into GSCM is severely lacking when it comes to small- and medium-sized enterprises (SMEs). The organisational management paper by Sharma (2000) requested that researchers give importance to environmental problems when conducting organisational studies. In the area of strategic management, Noci and Verganti (1999) noted that mainstream literature has focused on the examination of larger firms, and did not look into SMEs' features. Hitchens et al. (2004) stated that the issue is noticeable for SMEs, as they are further hindered by limited information and funding for green management practices. As a result, it is necessary to comprehend the reasons behind SMEs employing GSCM in their operations, and how they can best go about this task.

The remaining part of this study is structured as follows. The background of the study and literature review is discussed in Section 2. Section 3 represents the methodology. The findings and discussions are presented in Section 4. The recommendations and conclusions are in Section 5.

10.2 LITERATURE REVIEW

10.2.1 ENVIRONMENTAL MANAGEMENT

An EMS allows companies to conduct their business services and activities whilst minimising damage to the environment through a specific group of processes, values and management techniques (Feldman *et al.*, 1997). To support companies achieving the desired sustainable development, the International Standardisation Organisation (ISO) created the list of standards known as the ISO 14001. Ofori (2000) explains that, under ISO 14001, EMS represents one element within the total management system of a company. The EMS entails considerations such as the monitoring and evaluation of firms' environmental policies, the implementation and achievement of such policies, developmental resources and procedures, business practices, operational procedures, corporate duties, planning and organisational structure. The ISO 14001 is structured based on the 'plan-do-check-review' ideology and includes five cyclic stages of development and betterment. Ritchie and Hayes (1998) outline the various advantages of ISO 14001 as follows: environmental protection; cost reduction; improved market entry opportunities; proven regulatory compliance; more environmentally friendly performance; higher levels of customer satisfaction; stronger trust between consumers and the company; promotion of fair competition, influencing global trade; more knowledgeable and engaged employees; improved brand image; a better and more credible reputation for the firm.

10.2.2 SUPPLY CHAIN MANAGEMENT

Bachok *et al.* (2004) define the supply chain as a group of processes that facilitates the transference of knowledge, money, information and tangible products and services between supplier companies and end-users. Mentzer *et al.* (2001), however, explain that supply chain management (SCM) has various definitions and no overall consensus amongst those who have attempted to define the term. Some propose that the term SCM refers to a process that businesses use to associate suppliers to the company and the company to consumers, with the process being strategic in nature. Others have defined SCM not as a strategic element but as an art form that, when mastered, offers value to companies (Billington, 1999). Ofori (2000) explains that the benefit of SCM is that it provides the company with insight when forming business strategies and allows firms to identify opportunities and threats in the supply chain.

10.2.3 GREEN SUPPLY CHAIN MANAGEMENT

Today's companies are beginning to become much more aware of their duties towards the environment. Therefore, environmental management has started to emerge as a core strategic component of organisational performance. Zhu and Sarkis (2006) explain that the product lifecycle entails various stages at which the environment can be damaged, whether this is the collection of raw materials to create the product, the manufacturing process behind the product, or the discarding or recycling of the product. Sarkis (1995) points out that, for this reason, companies have begun to focus heavily on increasing their productivity and decreasing the damage they cause to the environment through GSCM in order to obtain stronger competitive advantages and profits. Rao (2005) adds that ISO 14000 (and the equivalent European Union's EMAS) have promoted the role of GSCM for some time. As Zhu and Sarkis (2004) explain, because GSCM entails somewhat modern research and practice areas (i.e. SCM and corporate environmental management), there is no absolute agreement on how to define it. Over the last two decades, various proposals have been made in solution to the lack of global definition for GSCM:

> The way in which innovation in supply chain management and industrial purchasing may be considered in the context of the environment
>
> The term "supply chain" describes the network of suppliers, distributors and consumers. It also includes transportation between the supplier and the consumer, as well as the final consumer... the environmental effects of the researching developing, manufacturing, storing, transporting, and using a product, as well as disposing of the product waste, must be considered

(Messelbeck and Whaley, 2000, p. 42)

There are various explanations about what are the GSCM activities and why organisations should consider engaging in the activities.

10.2.4   GREEN SUPPLY CHAIN MANAGEMENT IN SMEs AND ITS MOTIVATING FACTORS

GSCM research explores the impacts, motivators, practices and definitions involved in GSCM. GSCM practices are heavily based on monitoring (collecting and interpreting supplier information, putting supplier evaluation criteria into place, and assessing suppliers' environmental performance as well as the environmental performance of products) and collaboration (joint research with suppliers, information-sharing on environmental management, and supplier training incentives). Amongst the researchers who have examined these issues is Hall (2000) who explains that GSCM practices often start as a result of regulatory requirements or consumer/stakeholder demands.

According to Bowen *et al.* (2001), Theyel (2001) and Lippman (2001), firms then implement these practices through greater supplier compliance, senior management input, SCM capabilities, the use of multifunctional teams, and so on, making the best use of assets held both within the company and by other companies. Rao and Holt (2005) and Zhu and Sarkis (2004) have attempted to explore the ways in which GSCM practice and companies' financial and environmental performance are linked, often emphasising the companies rather than the suppliers. At the point of writing this research paper, the engagement of small-to-medium suppliers in large enterprises' GSCM schemes has rarely been covered.

Researchers such as Hall (2000) have explored the specific factors that impact the environmental management practices of SMEs (e.g. lack of funding, lack of technological assets, lack of human resources, fewer demands from stakeholders and less recognition of environmental problems). Others, such as Luken and Stares (2005) have presented case studies to explore environmental support schemes in relation to SMEs. In these studies, it is proposed that SMEs are more likely to focus on bettering their environmental performance when they are able to obtain support from relevant figureheads and leaders, or when they are under pressure from the supply chain. One study was more extensive and emphasised SMEs' implementation of clean technologies (Hitchens, 2003). The research that has been conducted on the environmental management practices of SMEs focuses, at least somewhat, on the motivations behind SMEs' adoption of environmental management strategies. In summary, these factors include the desire to enhance their environmental performance, pressure/assistance from external stakeholders, and regulatory requirements. This being said, the existing studies pay little attention to the supply chain. Consequently, the existing body of literature fails to examine the ways in which green supply chain schemes are impacted by SME suppliers. Furthermore, very few research papers discuss or test the factors that can motivate suppliers to become more involved in the green supply chain.

10.2.5 South Korean Green Supply Chain (GSC) Programmes

In the past, Korea was characterised by a determined drive towards achieving national economic growth, with environmental issues only entering Korea's social consciousness over the last 20 years. Consequently, as Lee and Rhee (2006) point out, this enabled a 500-fold increase to occur in the Korean economy for a period of some 30 years.

Environmental concerns were first raised in 1991 after the infamous phenol leakage incident, along with other serious events that caused damage to the environment. Furthermore, the Korean government began to focus on the supply chain's environmental impacts as a consequence of EU regulation (e.g. the Restriction of Hazardous Substances [RoHS] and Waste Electrical and Electronic Equipment Directives). Soon, Korean businesses began to follow suit (Lee, 2007). Foreign markets are hugely influential on the Korean economy, with many Korean businesses maintaining a competitive advantage as a direct result of their suppliers. For example, one report states that of all cars manufactured in Korea, around two-thirds were exported. Furthermore, it is reported that Korean businesses spend an average of 59% of their revenue on purchasing (Lee, K. H., 2008). Consequently, the supply chain became the target of an environmental management policy newly introduced by the Korean government, with SME suppliers becoming a major focus of the policy. This policy resulted in a national GSC scheme that began in 2003, which was initially designed to support SME suppliers in achieving better environmental performance. The government initiative was founded upon the North American Supplier Partnership for the Environment's set of standards and emphasised partnerships between leading large-size companies and the suppliers of these enterprises. Here, SCM is implemented as a key strategy to facilitate information-sharing between large purchasing companies and SME suppliers. The initiative is known as the GSC programme. The GSC programme is funded by participating firms and the government in South Korea and serves to offer all participating suppliers the opportunity to engage in extensive environmental support schemes. This offers benefits in the form of environmental data profiling, mass balance management, hazardous material analysis, and green procurement and management support from partnerships with leading experts and purchasing companies (Lee, S. Y., 2008). However, most of the SMEs still do not have the necessary background information, expertise, skills and economic or human resources to implement the necessary alterations in their firm (Lee, S. Y., 2008). Furthermore, it is often the case that their practices are concentrated on a specific aspect of the production process or the product that they want to change. As a result, SMEs usually have a restricted view of future innovation possibilities, and usually approach green supply chain problems (Nawrocka, 2008). After the EU parliament approved the proposed guidelines for Waste Electrical and Electronic Equipment (WEEE), Restriction of Hazardous Substances (RoHS), a number of Multinational Corporation (MNC) firms in the consumer products industry (such as Apple, Samsung, LG, Sharp, Lenovo, Sandisk and Nokia) have implemented 'green' standards in their SCM. End manufacturers use their buying power to convince suppliers to uphold a greater standard of environmental performance. Under the RoHS-compliance programme, a number of larger firms request that their suppliers verify parts and components compliance in order to ensure compliance of the final products (Cusack & Perrett, 2006). The majority of the suppliers along the supply chain are SMEs, and 4 out of every 5 global enterprises are categorised as SMEs (i.e. having less than 250 employees) (Moore & Manring, 2009). Considering that more than 80% of enterprises in the world are technically small and medium-size, there is even more reason to be concerned about the creation of a business case for the sustainability of SMEs and to encourage SME investments in sustainable business operations. When accounting for the stringent environmental requirements currently in place (and those in the future), SME business operations must be inspected in order to ascertain and solve environmental management and sustainability issues. To best comprehend the unique situation of implementing GSCM in SMEs, the strategic backgrounds of organisational structure, innovation capability and human resources, and cost-saving for GSCM are taken into account to provide additional clarity (Lee, K. H., 2009). The four strategic backgrounds will be explained with the chosen case study in the next chapter.

10.3   METHODOLOGY

This paper utilises literature review and qualitative research to achieve its research objectives. The inductive approach resulted in a set of raw data that connected the research objectives to the data.

Once condensed data was derived from the literature review, and the authors turned to qualitative research to further the study. Qualitative research formulates hypotheses using a variety of contextual inputs. In this instance, the authors used secondary sources – such as the internet, literature and articles – to inform the theory and objectives of this study.

There is a wealth of research concerning GSCM and sustainability, but there is a lack of research that specifically applies that information to practical decisions made by SMEs. This study attempts to fill the gap in research through descriptive and explorative means. The case study section of this research collected data through primary interviews and meetings supplemented by secondary data gleaned from company literature and presentations. In this way, both primary and secondary data methods were utilised in this study.

10.3.1 CASE STUDY: PEMTECH

PEMTHCH (Corp.) is a company that has been established since 1993, with capital expenditure of ₩ 615,370,000 (£ 352,296), and a sales figure of ₩ 28,851,890,000 (£ 16,397,300) per annum in 2014; and employing 134 staff. It is a company that has been selected by the small business administration in South Korea (SBA) as an outstanding small business (see Table 10.1). PEMTHCH's main industry of business is the manufacturing of construction materials, but also operates refrigerant storage warehousing, and red-clay housing business. This research facility is progressing with various research developments relating to the red-clay housing business, such as architectural design, boiler design, quality control of constructional materials along with research on function improvement, the system of selling products, and system of transportation. Operating in a variety of different fields of business, PEMTHCH's main business is civil engineering, specialising in construction solutions: it builds PEM retaining walls which is a protection measure for cut-slope from the field of slope stability. The main customers of PEMTECH include Samsung C&T, Hyundai Heavy Industries Co. Ltd, public enterprises (Korean Rail Network Authority, Korea Land & Housing Corporation, Korean-Water, etc.) and many regional construction management administrations.

10.3.2 GSCM's CHALLENGES IN PEMTECH AND CONSTRUCTION INDUSTRY

PEMTECH's P.C. panel manufacturing production process uses steel reinforcement, concrete and cement as raw materials, with cement being comprised of various components including lime, silica and alumina and being one of the simpler materials

---

TABLE 10.1
Drivers for Implementing the GSCM at PEMTECH

Internal Drivers	External Drivers
Reducing environmental costs	Complying with stringent government environmental regulations
Solving the recycling, reusing problem in P.C. panel process	Responding proactively for customers' green expectation
Saving the cost from the reusable machineries and materials	Increasing the green public image

used in construction. According to Occupational Safety and Health Administration (2008), cement production uses a high amount of energy and contributes as much as 10% of global CO_2 emissions: the world's third-largest generator of CO_2 (after transport and energy production).

> Moreover, PEMTECH's construction process leads to various waste issues, which can generally be categorised as direct waste or indirect waste. According to EPA (2002), indirect waste stems from low-quality workmanship, production waste, tailored specifications/dimensions and the nature of the materials. Direct waste, on the other hand, stems from trial-and-error, the inefficiency of the production plant, incorrect specifications, inappropriate usage, vandalism, theft, spillages, cutting, storage, delivery and transport. According to Envirowise (2005), two key incentives for improving the construction industry's resource efficiency are said to be: (i) the need for construction companies to show contractors and clients that they are in the continuous development of sustainable construction processes; and (ii) the increasing expense involved in construction products production and waste disposal. The Ministry of Environment (2013) in South Korea reports that the cost of waste disposal and construction products are being increased further by the following factors: (i) a £3-per-tonne annual rise in landfill tax, which was predicted to be £27 per-tonne or more by 2018; (ii) an increase in the cost of waste disposal due to stricter regulations and less space in landfills, and (iii) the aggregates levy and other taxes and levies impacting primary materials. Therefore, waste management and reusing processes are critical issues for GSCM at PEMTECH.

10.4   FINDINGS AND DISCUSSIONS

This paper has explored the implementation of GSCM amongst Korean SMEs within the construction industry from a variety of angles, including the exploration of GSC practices and activities in order to assess the development and success in this area. Amongst the core considerations behind SMEs' proactive or reactive adoption of GSCM is the driving forces behind the adoption of this corporate management style. The motivational factors to implement GSC practices are the cause of SMEs move toward GSCM. This paper has revealed that PEMTECH's adoption of ISO 14001 stemmed from the need to meet a number of demands arising outside of the company environment. As part of this, the company began to examine its competitive strengths in the market, regulatory issues, absence of waste management processes and production process. As shown in Table 10.1, the company essentially decided to move towards greener management due to demand from the public and the need to adhere to environmental legislation. As a consequence of increasingly strict environmental regulations, PEMTECH implemented a strict standard for emissions.

As supported by Perez-Sanchez et al. (2003), an analysis of smaller firms revealed that these two factors (public pressure and environmental regulations) are the primarily motivators behind companies' adoption of environmental management strategies. Of course, all companies strive to adhere to fundamental environmental legislation. However, as Azzone et al. (1997) point out, SMEs encounter greater impacts from environmental regulations due to their possession of fewer resources than large companies. Thus, since SMEs often lack adequate funds, and because the adoption of environmental technologies is complicated, this tends to mean that SMEs face greater challenges in this regard.

It is argued that SMEs' lower level of experience and technological complexity leave them in a weaker position than large companies (Lee, K. H., 2008, 2009). All companies must deal with various legal, managerial, technological and political issues when facing legislation. Since numerous norms exist, this results in a more complicated process, which brings additional challenges in dealing with the interaction of managerial processes, manufacturing processes, environmental technologies and regulators. This indicates that companies must enact a greater number of activities the more they are subject to restrictions from environmental legislation. Thus, SMEs might meet requirements by changing their organisational procedures and increase efficiency by selecting the most appropriate environmental technologies.

10.5 RECOMMENDATIONS AND CONCLUSIONS

The ability of companies to remain competitive in the market will greatly depend on sustainability factors in the future. Thus, it is essential to gain greater insight into the relationship between GSCM and competitive advantage in order to ascertain whether businesses stand to gain from investing in sustainability and environmental management. As Del Brío and Junquera (2003) point out, business and management research must immediately begin to focus more on the competitive impacts of environmental regulations for SMEs, especially since very few studies have explored the relationship between environmental management, GSCM and SME performance.

Whilst it can be argued that there are no valid reasons for avoiding the adoption of GSCM, companies continue to pose numerous reasons for their hesitance. Essentially, companies must consider whether or not they are prepared to contribute in a positive way by committing to navigate the difficulties that arise in making these necessary changes. It is clear that there is a strong case for companies to become implementing GSC activities successfully and that they stand to gain competitive advantages from doing so. However, unless managers are driven by appropriate motivators, it is unlikely that every SME will take steps to improve its environmental performance in any tangible way. Environmental legislation has been highlighted in this research study as a crucial motivator, but the extent to which it can motivate SMEs to become greener depends on the nature of what is required by the company in order to adhere to regulatory requirements. Additionally, the research findings suggest that companies can transform various aspects of management practice to increase their environmental performance. Specifically, this can be achieved if SMEs focus on competitive advantage, cost savings, human resource development, innovation, learning and the GSCM's activities (Green production, Green purchasing, Green logistic, Reuse). The research findings obtained in this study have successfully shed light on the options available to SMEs when making changes to their organisational strategies in order to adopt GSCM practices. Further exploration of the topic is highly encouraged, particularly in the context of other industries and a broader variety of enterprises in order to overcome some of the limitations present in the current study. Additionally, in exploring the difference between reactivity and proactivity in organisational strategy, it could be useful for future studies to apply quantitative methods to a single industry. It could also be useful for other researchers to explore ways to assess GSCM implementation through the establishment of dedicated strategies and guidelines. Being able to evaluate the level of adoption within a company through the use of cultivation indicators would be greatly beneficial, as would the ability to evaluate a company's immersion in GSCM.

REFERENCES

Azzone, G., Bianchi, R., Mauri, R., & Noci, G. (1997). Defining operating environmental strategies: Programmes and plans within Italian industries. *Environmental Management and Health*, 8(1), 4–19.

Bachok, S., Al-Habshi, S. M. S., Jaafar, S., & Baharudin, H. (2004). Construction supply chain management and coordinated design drawings: An outlook of the construction industry and sustainable urban planning. 9th International Symposition on Planning & IT, Vienna, Februray, 67–84.

Billington, C. (1999). The language of supply chains. *Supply Chain Management Review*, 111(2), 86. Available at: <http://e3associates.com/files/Article%20%20The%20Language%20of%20Supply%20Chain.pdf> (Accessed *August 2015*)

Bowen, F. E., Cousins, P. D., Lamming, R. C., & Farukt, A. C. (2001). The role of supply management capabilities in green supply. *Production and Operations Management*, 10(2), 174–189.

Bowersox, D. J., & Closs, D. J. (1996). *Logistics management*. New York: McGraw Hill, 66–82.

Cusack, P., & Perrett, T. (2006). The EU RoHS directive and its implications for the plastics industry. *Plastics, Additives and Compounding*, 8(3), 46–49.

Del Brío, J. Á., & Junquera, B. (2003). A review of the literature on environmental innovation management in SMEs: Implications for public policies. *Technovation*, 23(12), 939–948.

Envirowise, (2005). Saving money and raw materials by reducing waste in construction: Case studies. London: Envirowise. Available at <http://www.gloucestershire.gov.uk/media/adobe_acrobat/2/1/Envirowise%20Case%20Studies.pdf> (*Accessed August 2015*)

EPA (2002). What is integrated solid waste management? Environmental protection agency, Available at <http://epa.gov/climatechange/wycd/waste/downloads/overview.pdf> (*Accessed September 2015*)

Feldman, S. J., Soyka, P. A., & Ameer, P. G. (1997). Does improving a firm's environmental management system and environmental performance result in a higher stock price? *The Journal of Investing*, 6(4), 87–97.

Green, K., Morton, B., &New, S. (1996). Purchasing and environmental management: Interaction, policies and opportunities. *Business Strategy and the Environment*, 5, 188–197.

Hall, J. (2000). Environmental supply chain dynamics. *Journal of Cleaner Production*, 8(6), 455–471.

Hart, S. L. (2005). Innovation, creative destruction and sustainability. *Research-Technology Management*, 48(5), 21–27.

Hitchens, D. (2003). *Small and medium sized companies in Europe: Environmental performance, competitiveness and management: International EU case studies*. Springer Science & Business Media.

Hitchens, D., Clausen, J., Trainor, M., Keil, M., & Thankappan, S. (2004), Competitiveness, environmental performance and management of SMEs. *Greener Management International*, 44, 45–57.

Hutchinson, C. (1996). Corporate strategy and the environment. *Business and the Environment*, 85–104.

Kannan, D., de Sousa Jabbour, A. B. L., & Jabbour, C. J. C. (2014). Selecting green suppliers based on GSCM practices: Using fuzzy TOPSIS applied to a Brazilian electronics company. *European Journal of Operational Research, 233*(2), 432–447.

Lippman, S. (2001). Supply chain environmental management. *Environmental Quality Management, 11*(2), 11–14.

Lee, K-H. (2007). Corporate social responsiveness in the Korean electronics industry. *Corporate Social Responsibility and Environmental Management, 14*(4), 219–230.

Lee, K-H. (2008), Corporate environmental management and practices of SMEs: The case of Korean manufacturing industry. *Journal of Sustainable Management, 8*(1), 73–86.

Lee, K. H. (2009). Why and how to adopt green management into business organizations? The case study of Korean SMEs in manufacturing industry. *Management Decision, 47*(7), 1101–1121.

Lee, S. Y. (2008). Drivers for the participation of small and medium-sized suppliers in green supply chain initiatives. *Supply Chain Management: An International Journal, 13*(3), 185–198.

Lee, S. Y., & Klassen, R. D. (2008). Drivers and enablers that foster environmental management capabilities in small-and medium-sized suppliers in supply chains. *Production and Operations Management, 17*(6), 573–586.

Lee, S. Y., & Rhee, S. K. (2007). The change in corporate environmental strategies: A longitudinal empirical study. *Management Decision, 45*(2), 196–216.

LG Electronics (2014), Sustainability report. Available at <http://www.lg.com/global/sustainability/communications/sustainability-reports> (*Accessed August 2015*)

Luken, R., & Stares, R. (2005). Small business responsibility in developing countries: A threat or an opportunity? *Business Strategy and the Environment, 14*(1), 38–53.

McKinsey & Company (2008), How companies think about climate change: A McKinsey global survey, The McKinsey Quarterly, February, p. 18. Available at <http://www.nyu.edu/intercep/lapietra/ClimateChangeAttitudes.pdf> (*Accessed September 2015*)

Mentzer, J. T., DeWitt, W., Keebler, J. S., Min, S., Nix, N. W., Smith, C. D., & Zacharia, Z. G. (2001). Defining supply chain management. *Journal of Business Logistics, 22*(2), 1–25.

Messelbeck, J., & Whaley, M. (2000). Greening the health care supply chain: Triggers of change, models for success. *Corporate Environmental Strategy, 6*(1), 39–45.

Min, H., & Galle, W. P. (2001). Green purchasing practices of US firms. *International Journal of Operations & Production Management, 21*(9), 1222–1238.

Min, H., & Galle, W. P. (1997). Green purchasing strategies: Trends and implications. *International Journal of Purchasing and Materials Management, 33*(3), 10–17.

Ministry of Environment (2013). South Korea's waste management policies: Information note, Ministry of Environment in South Korea. Available at <http://www.legco.gov.hk/yr12-13/english/sec/library/1213inc04-e.pdf > (*Accessed August 2015*)

Moore, S. B., & Manring, S. L. (2009). Strategy development in small and medium sized enterprises for sustainability and increased value creation. *Journal of Cleaner Production, 17*(2), 276–282.

Narasimhan, R., & Carter, J. R. (1998). *Environmental supply chain management.* Center for Advanced Purchasing Studies.

Nawrocka, D. (2008). Environmental supply chain management, ISO 14001 and RoHS. How are small companies in the electronics sector managing? *Corporate Social Responsibility and Environmental Management, 15*(6), 349–360.

Noci, G., & Verganti, R. (1999). Managing 'green' product innovation in small firms. *R&D Management, 29*(1), 3–15.

Occupational Safety and Health Administration (2008). Preventing Skin Problems form Working with Portland Cement, U.S. Department of Labor. Available at <https://www.osha.gov/Publications/OSHA-3351-portland-cement.pdf > (*Accessed August 2015*)

Ofori, G. (2000). Greening the construction supply chain in Singapore. *European Journal of Purchasing & Supply Management, 6*(3), 195–206.

Ojo, E., Akinlabi, E. T., & Mbohwa, C. (2012), Benefits of green supply chain management in construction firms – A review. 2nd Nelson Mandela Metropolitan University, Construction Management Conference.

Palmer, K., Oates, W., & Portney, P. (1995). Tightening environmental standards: The benefit-cost or the no-cost paradigm?. *Journal of Economic Perspectives, 9*(4), 119–132.

Perez-Sanchez, D., Barton, J. R., & Bower, D. (2003). Implementing environmental management in SMEs. *Corporate Social Responsibility and Environmental Management, 10*(2), 67–77.

Porter, M. E., & Van der Linde, C. (1995). Green and competitive: Ending the stalemate. *Harvard Business Review, 73*(5), 120–134.

Rao, P., & Holt, D. (2005). Do green supply chains lead to competitiveness and economic performance? *International Journal of Operations & Production Management, 25*(9), 898–916.

Rao, P. (2005). The greening of suppliers—In the South East Asian context. *Journal of Cleaner Production, 13*(9), 935–945.

Ritchie, I., & Hayes, W. (1998). *A Guide to the Implementation of the ISO 14000 Series on Environmental Management*. New Jersey: Prentice Hall.

Samsung Electronics (2014). Sustainability report. Available at http://www.samsung.com/us/aboutsamsung/investor_relations/corporate_governance/corporatesocialresponsibility/downloads/2014sustainabilityreport.pdf> (*Accessed August 2015*)

Sarkis, J. (1995). Manufacturing strategy and environmental consciousness. *Technovation, 15*(2), 79–97.

Sarkis, J. (2003). A strategic decision framework for green supply chain management. *Journal of Cleaner Production, 11*, 397–409.

Sbihi, A., & Eglese, R. W. (2010). Combinatorial optimization and green logistics. *Annals of Operations Research, 175*(1), 159–175.

Sharma, S. (2000). Managerial interpretations and organizational context as predictors of corporate choice of environmental strategy. *Academy of Management Journal, 43*(4), 681–697.

Shekari, H., Shirazi, S., Afshari, M., & Veyseh, S. (2011). Analyzing the key factors affecting the green supply chain management: A case study of steel industry. *Management Science Letters, 1*(4), 541–550.

Srivastava, S. K. (2007). Green supply-chain management: A state-of-the-art literature review. *International Journal of Management Reviews, 9*(1), 53–80.

Testa, F., & Iraldo, F. (2010). Shadows and lights of GSCM (Green Supply Chain Management): Determinants and effects of these practices based on a multi-national study. *Journal of Cleaner Production, 18*(10), 953–962.

Theyel, G. (2001). Customer and supplier relations for environmental performance. *Greener Management International, (35)*, 61–69.

Yin, R. K. (2013). *Case study research: Design and methods*. Sage publications.

Zhu, Q., & Sarkis, J. (2004). Relationships between operational practices and performance among early adopters of green supply chain management practices in Chinese manufacturing enterprises. *Journal of Operations Management, 22*(3), 265–289.

Zhu, Q., & Sarkis, J. (2006). An inter-sectoral comparison of green supply chain management in China: Drivers and practices. *Journal of Cleaner Production, 14*(5), 472–486.

Zhu, Q., Sarkis, J., & Lai, K. H. (2007). Initiatives and outcomes of green supply chain management implementation by Chinese manufacturers. *Journal of Environmental Management, 85*(1), 179–189.

Zsidisin, G. A., & Siferd, S. P. (2001). Environmental purchasing: A framework for theory development. *European Journal of Purchasing & Supply Management, 7*(1), 61–73.

11 Drivers for Adoption of Green Logistics as a Means to Achieve Circular Economy by Organized Retail Sector

R.A. Dakshina Murthy
Prin. L.N. Welingkar Institute of Management
Development & Research
Bangalore, Karnataka, India

Leena James
Deanery of Commerce & Management,
Christ Deemed University, Bangalore, Karnataka, India

CONTENTS

11.1 INTRODUCTION

Logistics activity include movement, storage and handling of goods along the supply chain. The prime objective for the logistic players has been to make logistics operation commercially viable. However, logistics operation cause adverse impact on the environment and the logistics players hardly have ever paid attention to this aspect. Green Logistics (GL) aims to perform all the logistics-related operation in a more sustainable manner by reducing the negative impact of logistics on the environment. GL is the concept that links between resources and products and products to the consumers. GL helps in closing the loop in the circular economy (CE).

From the Indian Logistics Operation context, the logistics activities are projected to release 680 million tonnes of CO_2e (Carbon Dioxide Equivalent) by the year **2030** as there could be seven-fold increase in vehicle fleet to about 380 million vehicles. Logistics play an important role for the Retail & e-tail sector and enable them to be very effective operationally. Those who are in e-commerce business employ more logistics services to ensure faster delivery and better service; but they tend to ignore the adverse effect of the same on the environment. Due to ever increasing economic activity year on year, and the usage of the logistics services, the Green House Gas (GHG) emissions are projected to double over the next decade from the present level of 7–8%. GL adoption is at nascent stage and the various stakeholders in logistics business are needed to be made aware of GL and should be driven into adopting GL.

Logistics operation is not limited to only transportation but it entails other operational activities such as packaging/re-packaging, storage, warehousing, inventory management and optimum use of all resources including energy. Ensuring sustainability in operational activities of a typical retail not only reduce GHG emissions but also leads to achieve CE through the creation of robust and very efficient supply chain for which logistics operation is vital. Organizations dealing with retail business are required to strategize to achieve set business goals without compromising on the sustainability aspects of business. Stahel, one of the founding fathers of the concept of CE, states that CE entails optimum utilization of all resources (Stahel, 2016). CE would benefit environment, economy and society (Fonseca et al., 2018). The concept of CE bring balance among business goals and environmental sustainability and as well benefit the society with emphasis on economic and social dimensions (Murray, 2017; Naustdalslid, 2014). CE drives a point that both a healthy economy and a healthy environment can coexist (Geng et al., 2008). Application of CE aims to minimize the adverse effects of operation through reducing and minimizing the release of GHG to the environment. CE incorporates concepts such as 3R, cleaner manufacturing, industrial ecology, Green SCM, GL etc., to mitigate the adverse effects on environment and GL is considered as a pre-requisite for the development of CE in the logistics sector (Roberta et al., 2018). To survive in the global market, implementation of sustainable supply chain management practices and integrating with pollution prevention strategies to minimize environmental pollution are crucial (Kumar et al., 2020).

The aim of this study is to ascertain level of awareness of GL, determine the drivers for adoption of GL by Organized Retail Sector and indicate measures to achieve CE that facilitates establishment of sustainable operational practices.

11.2   REVIEW OF LITERATURE

Logistics industry is one of the major contributor to environmental degradation due to heavy traffic, poor choice of transport modes and lack of infrastructure (Whitelegg, 1993). The members of the Logistics and Distribution Management in the United Kingdom indicated that the Government Legislation, Employees and Customers involvement and the influence by the lobby groups are the key factors for adoption of GL (Szymankiewicz, 1993). Distance traversed by a material or product to finally reach the consumer is known as food mile and food mile increases if the suppliers are far away (Boge, 1994). Federal Express employed single warehouse to serve the stores located across the country that ensured very high levels of service (Aron, 1994). International Journal of Physical Distribution & Logistics Management came up with a special edition on 'environmental aspects of logistics', which highlighted about fewer shipments, more direct routes, shorter movements, lesser handling facilitates GL adoption but it works at the cost of trade-offs among responsiveness, cost and quality, delivery time and customer satisfaction (Wu and Dunn, 1995).

Proper disposal of waste is utmost essential to reduce pollution of the environment especially that of air and water (Murphy *et al.*, 1994, 1995,1996). The factors such as government regulations, meeting the social needs, de-risking from environmental damage lawsuits, which reduces the cost, benchmarking with the competition leading to better profitability are the key parameters driving the adoption of GL (Caputo & Mininno, 1996). Vertical integration interventions with the logistics operation that may have positive effect on the overall logistics performance and companies that emphasizes GL with good integration with suppliers and logistics service providers can obtain best route scheduling (Millighan, 2000).

Due to enhanced customer awareness about environmental aspects, Green Supply Chain is assuming importance (Beamon, 1999). Study related to multi-objective optimization provided several set of solutions for system enhancement and importance to environmental aspect determines the logistic design and solutions with a view to achieve the objectives of sustainability (Azapagic & Clift, 1999). There have been efforts by various organization to make the supply chain green and sustainable and these efforts are also applied to the various activities coming under logistics management (Skjoett-Larsen, 2000; Bacallan, 2000). Perceptible change noticed in people with regard to environmental concern and also noticed growing importance of environmental sustainability across the globe (Srivastava, 2007).

The materials used all along the supply chain must be environmentally friendly and the processes should be clean and green (Corbett & Kleindorfer, 2003). If the demand forecast accuracy improves, inventories can be optimized at various points along the supply chain and thus greener logistics solution can be employed (Rothenberg, 2001). The Logistics Service Providers took many initiatives to enhance their service value (Murphy & Daley, 2001). The role of logistics service providers is to design and offer GL solutions for the clients based on the specific requirements (Murphy & Poist, 2003; Sarkis *et al.*, 2004). The extent of volume of new vehicles getting onto the road are so high that it outweighs any improvements by vehicle manufacturers with regard to emission levels with greener technologies as it has failed to catch up with the growing volume of logistics requirements (Aronsson & Brodin, 2006). Lower environment footprint observed in case of online e-tailers as compared to conventional retailers (Smithers, 2007). Factory Gate Pricing with better coordination among the primary transporters resulted in reduction in CO_2e emissions and back-loading strategy adopted through better consolidation and coordination among various stakeholders can reduce the vehicle km per annum used by retail units (Mckinnon & Ge, 2004; Potter *et al.*, 2007). Vehicle routing through networking and optimization of trips through proper planning of transport requirements and avoiding congested traffic routes will cause less GHG emissions (McKinnon, 2011; Svensson, 2007).

But very low levels of Green Supply Chain Practices were reported (Dakshina Murthy & James, 2014). Only 14% of the organized retail sector were aware of GL concepts and very few have initiated measures towards environmental friendly logistic activities (Kumar, S. *et al.*, 2012). Organizations are aware of green concepts but are yet to implement the practices related to sustainable operation (Kumar, N. *et al.*, 2012). In any typical logistics operation, the main activity is the transport part and in certain countries it is predominantly road transport Matthias, 2016). Shifting mode of transport from roadways to railway is key to reduce the GHG emissions (Islam *et al.*, 2012). Barriers for implementing GL are the high investment, lack of will and lack of infrastructure support from the Government (El-Berishy, N. 2013).

Embedding CE principles in SCM lead to development of circular supply chain through incorporation of concept of leasing instead of owning, inclusion of public and private procurement agencies and management of waste for reuse and recycling purposes (Zink & Geyer, 2017). GL is one of the basic necessity for development of CE as it ensures smooth flow of materials into the system (Oksana & Agniezka, 2019). GL aims to reduce the ecological and energy footprint in the distribution process through lower emissions, reduced energy consumption and has a positive social impact as well (Su-Yol Lee & Klassen, 2008; Sbihi & Eglese, 2009). Supply Chain and Logistics, which are the back bone of retail, are the inevitable choices for making the business sustainable (Park, 2010). In this regard, the concept of GL directly correlates CE with logistics and CE and GL are directly linked (Guihua, 2013).

Greening of Logistics operation is key to achieve the sustainable business goals of an organization (Zou, 2015; Rong, 2014). Logistics operation based on the concepts of CE requires that the design of logistics solution is to be aimed at creating ecological balance with optimum utilization of resources (and logistics solution based on GL are a pre-requisite for achieving CE goals in the logistics sector (Zheng & Zheng, 2010). Social responsibility can be leveraged as the driver for establishing the sustainable supply chain (Kumar, *et al.*, 2019) and supportive government policies have a critical impact on the adoption of Industry 4:0 practices, which enable sustainable business practices (Sunil, L et al., 2020).

The adoption of GL by the retail industry is quintessential and the question to pose now is: 'The efficiency and effectiveness of logistics industry, which has seen tremendous growth over the last decade; are they equally compatible with the environment'?

11.2.1 RESEARCH GAPS

Logistics operation is one the prime operation involved with the retail business and the retail industry at present is more concerned about achieving the commercial viability while adverse effects of logistics on the environment is often overlooked. Awareness about GL and adoption of GL by organized retail sector is of paramount importance considering that environmental sustainability is also one of the goals of the industry apart from achieving the commercial success. Retail players, therefore, need to understand the drivers for adoption of GL for quicker adoption of GL practices leading to achieving the CE.

Sustainable procurement initiatives implementation decisions are crucial because they affect the effectiveness and responsiveness of supply chain activities. SPIs implementation is not only addressing economic goals of the firm but also incorporate social and ecological concerns. Therefore, focal firms have to choose and evaluate their partners not only considering traditional selection criteria but need to add more SPIs implementation based criteria to achieve success.

11.3 PROBLEM DEFINITION

Organized retail sector has witnessed double digit growth over the last decade and projected to achieve exponential growth in the coming years. Environmental sustainability is assuming importance and the operational practices have to be specifically put in place with an objective to achieve operational efficiency with least impact on the environment. Logistics operation have adverse effect on the environment and adoption of green practices in logistics is essential to reduce its impact on the environment.

11.4 SOLUTION METHODOLOGY

Organized retail players operating in different formats such as Convenient Stores, Specialty Stores, Departmental stores, Super and Hyper Market stores dealing with Grocery, Food & Beverage, Apparel, consumer electronics goods, FMCG and few e-tail organizations were included as representatives of the population being studied. Simple random sampling method employed and primary data so collected subjected to the Statistical Analysis using SPSS software. Regression analysis tests carried out to test the hypothesis. Structural Equation Modeling using AMOS Software developed to present the theoretical framework for the research and the measurement model is indicated in the Figure 11.1. Adoption of GL and sustainable operational practices by organized retail sector leading to achieving the CE is discussed.

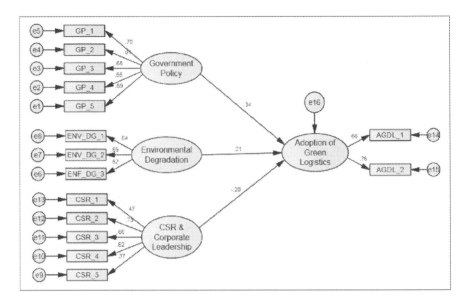

FIGURE 11.1 Measurement model for the influence of government policy, environmental degradation and CSR and corporate leadership on adoption of green logistics dimension.

Hypothesis formulated for the study are:

H_1: *Government Policy and Regulations* has significant influence towards Sustained *Adoption of Green Logistics*.

H_2: Environmental Degradation has significant influence towards S*ustained Adoption of Green Logistics*.

H_3: CSR & *Corporate Leadership in Sustainability* has significant influence towards S*ustained Adoption of Green Logistics*.

11.5   ANALYSIS OF RESULTS AND FINDINGS

11.5.1   TEST OF ASSUMPTIONS

Table 11.1 shows the Normality and Collinearity test results for the variables. Kurtosis and asymmetry values indicate normal univariate distribution as all the data lie between -2 and +2 (George & Mallery, 2010). Thus, Normality holds as the values are within the specified range of skewness and Kurtosis for all the items (Table 11.1). Hence, the assumption of normality holds for item under components of *Government Policy & Regulations, Environmental Degradation, CSR & Corporate Leadership in Sustainability.* The Variance Inflation Factor (VIF) and tolerance values of all the items are less than the acceptable value of 5 and 1, respectively, indicating very low collinearity and there is no possibility of inflated variance of the items. To assess *autocorrelation,* Durbin Watson statistics was calculated and the statistic values in the range of 1.5–2.5 are relatively normal (Field, 2009). Durbin Watson statistics for all dimensions are within the acceptable value.

TABLE 11.1

Normality and Collinearity Test Results by Item Wise Under All Dimensions

	Normality Test		Collinearity and Autocorrelation Statistics			
Item Label	Skewness	Kurtosis	Tolerance	VIF	Condition Indices	Durbin-Watson Statistic
		Government Policy and Regulations				
GP_1	-0.732	0.627	0.777	1.286	8.554	1.867
GP_2	-0.639	-0.520	0.821	1.218	11.254	
GP_3	-0.114	-1.001	0.883	1.132	12.998	
GP_4	-0.401	-0.675	0.821	1.218	17.095	
GP_5	-0.670	0.842	0.872	1.147	19.053	
		Environmental Degradation				
ENV_DG_1	-0.798	0.395	0.756	1.323	9.140	1.982
ENV_DG_2	-0.657	-0.186	0.753	1.329	12.478	
ENV_DG_3	-0.499	-0.604	0.932	1.073	14.439	
		CSR and Corporate Leadership in Sustainability				
CSR_1	-0.196	-1.053	0.873	1.146	9.211	1.961
CSR_2	-0.324	-0.720	0.701	1.426	9.916	
CSR_3	-0.307	-0.854	0.742	1.347	11.968	
CSR_4	-0.394	-0.779	0.771	1.298	12.967	
CSR_5	-0.024	-1.205	0.869	1.150	14.962	

11.5.2 CONVERGENT VALIDITY

It can be seen from Table 11.2 that the standardized loadings are greater than 0.60 in almost all the parameters for each construct. When items have factor loadings in the range 0.3 and 0.4, it is inferred that the items have achieved a minimum level of convergence and if the loadings are in the range 0.4 and 0.5, then items are considered as important (Hair *et al.*, 1995). When the items have loadings 0.50 and above, then they are considered to be significant (Ndubisi, 2006).

Speaking about the reliability factor, it is observed from Table 11.2 that the composite reliability and Cronbach alpha values are higher than the required value of 0.6 for *GOVERNMENT POLICY & REGULATIONS, ENVIRONMENTAL DEGRADATION* and *CSR & CORPORATE LEADERSHIP IN SUSTAINABILITY*. We can conclude that for majority of the items, which are grouped to a particular construct, completely converge to its respective sub dimensions and all the items can be considered for the study.

11.5.3 REGRESSION TEST RESULT

The regression results are provided in Table 11.3. Accordingly, it is observed that *GOVERNMENT POLICY & REGULATIONS* has a positive and significant influence on *Adoption of GL* dimension ($\beta = 0.344$; CR= 3.417, $p = 0.000$, $p < 0.05$).

TABLE 11.2
Reliability and Item Loadings of Manifest Indicators

Latent Variable	Indicators	Standardized Loadings (β)	Composite Reliability	Cronbach Alfa	Average Variance Explained (AVE)
Government	GP_1	0.702	0.783	0.784	0.421
Policy and	GP_2	0.611			
Regulations	GP_3	0.679			
	GP_4	0.551			
	GP_5	0.690			
Environmental	ENV_DG_1	0.522	0.732	0.739	0.486
Degradation	ENV_DG_2	0.840			
	ENV_DG_3	0.692			
CSR and	CSR_1	0.474	0.725	0.737	0.359
Corporate	CSR_2	0.781			
Leadership in	CSR_3	0.663			
Sustainability	CSR_4	0.621			
	CSR_5	0.368			
Adoption of	ADOPTION OF GREEN LOGISTICS_1	0.658	0.670	0.671	0.505
Green Logistics	ADOPTION OF GREEN LOGISTICS_2	0.760			

TABLE 11.3
Direct Effect of Research Model: Standardized Regression Weights for the Influence of Government Policy & Regulations, Environmental Degradation and CSR & Corporate Leadership in Sustainability on Adoption of Green Logistics Dimension

Relationships Between Variables Under Study			Standard Estimate	S.E.	C.R.	P-value
Adoption of Green Logistics	<---	Government Policy & Regulations	0.344	0.109	3.417	0.000*
Adoption of Green Logistics	<---	Environmental Degradation	0.208	0.128	2.263	0.024*
Adoption of Green Logistics	<---	CSR & Corporate Leadership in Sustainability	0.199	0.187	2.075	0.038*

* Significance at 5% level.

Thus, H_1 could be fully asserted. It can be implied that, for one-unit increase in the rating scale of agreement on *GOVERNMENT POLICY & REGULATIONS* construct, there can be 0.344 times (more than one third times) increase in *Adoption of GL* dimension.

Similarly, *ENVIRONMENTAL DEGRADATION* has a positive and significant influence on *Adoption of GL* dimension ($\beta = 0.208$; CR= 2.263, $p = 0.024$, $p < 0.05$). Thus, H_1 could be fully asserted. It can be implied that, for one-unit increase in the rating scale of agreement on *ENVIRONMENTAL DEGRADATION* construct, there can be 0.208 times (nearly one fifth times) increase in *Adoption of GL* dimension.

Finally, *CSR & CORPORATE LEADERSHIP IN SUSTAINABILITY* has a positive and significant influence on *Adoption of GL* dimension ($\beta = 0.199$; CR= 2.075, $p = 0.038$, $p < 0.05$). Thus, H_1 could be fully asserted. It can be implied that, for one-unit increase in the rating scale of agreement on *CSR & CORPORATE LEADERSHIP IN SUSTAINABILITY* construct, there can be 0.199 times (nearly one fifth times) increase in *Adoption of GL* dimension.

11.5.4   Goodness of Fit

The measurement model as developed using Structural Equation Modeling was tested for Goodness of Fit of the model and Table 11.4 indicates the various measures used for ascertaining the best fit of the model.

From Table 11.4, it is observed that the Chi-square/*df* (χ^2/df) is 1.864 (which is less than 3). To minimize the influence of sample size on the Chi-Square model, normed chi-square (χ^2/df) is used as a statistic and value below 3 is considered as suggesting good fit and the value less than 2 suggest very good fit of the model.

Goodness of Fit index (GFI) value obtained is 0.932 and it is more than 0.9, which suggest very good fit of the measurement model. As GFI is very sensitive to the

TABLE 11.4
Goodness-of-Fit and Incremental Indices of Measurement Model Influencing *Sustained Adoption of Green Logistics*

Fit Indices	Accepted Value	Model Value
Absolute Fit Measures		
χ^2 (Chi-square)		115.592
df (Degrees of Freedom)		62
Chi-square/*df* (χ^2/df)	< 3	1.864
GFI (Goodness of Fit Index)	> 0.90	0.932
RMSEA (Root Mean Square Error of Approximation)	< 0.10	0.060
AGFI (Adjusted Goodness of Fit Index)	> 0.90	0.900
Parsimony Fit Measures		
PCFI (Parsimony Comparative of Fit Index)	> 0.50	0.694
PNFI (Parsimony Normed Fit Index)	> 0.50	0.612

sample size, we can consider taking the values of Adjusted GFI (AGFI), which adjusts the GFI taking into consideration the degrees of freedom. AGFI value obtained is 0.9 and it suggest very good fit of the measurement model.

Root Mean Square Error of Approximation, which is most commonly used index for judging the goodness of fit shows a value of 0.06 and this value is well below the stringent cut off 0.07. Normally RMSEA parameter is predominantly used for assessing the goodness of fit of the structural model of SEM and for the aspect of influencing factors for sustained adoption of GL, the value of RMSEA supports the model indicating very good fit.

Considering that many of the parameters of the goodness of fit and incremental indices as shown above suggest that the measurement model is overall acceptable fit and the model can be considered as Over-Identified.

11.5.5 ANOVA Analysis

Retail players considered for study are dealing with different categories of products such as Food, Grocery, Consumer Electronic Goods, Apparel and General merchandise. In order to know and understand if there are any differences in the opinion among players dealing with different categories of products, hypothesis is formulated for each construct to see statistically any significant difference are there among retail players dealing with different categories of products. One-way Analysis of Variance (ANOVA) test conducted to see if there exist significant difference among various types of retailers with regard to variables being studied.

H_{01}: There is no significant difference in mean rating scores on level of agreement on *GOVERNMENT POLICY & REGULATIONS, ENVIRONMENTAL DEGRADATION* and *CSR & CORPORATE LEADERSHIP IN SUSTAINABILITY* across three types of Retail Outlets *(Apparel, General & Grocery, Consumer Electronic Goods).*

Tables 11.5–11.8 indicate the summed up values for each dimension based on the one-way ANOVA analysis.

It is seen from one-way ANOVA result (Table 11.5) that there is no statistically significant difference in overall mean score of *GOVERNMENT POLICY & REGULATIONS* $[F_{(2,240)} = 0.492, p = 0.612, p > 0.05]$ dimension among three types of Retail Outlets. F value is 0.492 with degrees of freedom of 2 and 240, and the *p*-value is greater than 0.05 at 5% significance, strongly indicating no difference between the opinions among retailers dealing with different categories of products with regard to the importance of government policy and regulations as regards the sustained adoption of GL. Hence, we accept the null hypothesis. In other words, mean rating score on degree of agreement with regard to *GOVERNMENT POLICY & REGULATIONS* dimension does not differ significantly between respondents working exclusively in *Apparel Retail* outlets and respondents working in *General & Grocery* and respondents working in *Consumer Electronic Goods* Retail Outlets.

TABLE 11.5
One-Way ANOVA Result Between Type of Outlets and *GOVERNMENT POLICY & REGULATIONS* Dimension

| | GOVERNMENT POLICY & REGULATIONS | | | | |
	Sum of Squares	df	Mean Square	F	p-value
Between Groups	12.242	2	6.121	0.492	0.612
Within Groups	2985.544	240	12.44		
Total	2997.786	242			

TABLE 11.6
One-Way ANOVA Result Between Type of Outlets and *ENVIRONMENTAL DEGRADATION* Dimension

| | ENVIRONMENTAL DEGRADATION | | | | |
	Sum of Squares	df	Mean Square	F	p-value
Between Groups	5.193	2	2.597	0.237	0.855
Within Groups	2620.472	240	10.918		
Total	2625.665	242			

TABLE 11.7
One-Way ANOVA Result Between Type of Outlets and *CSR & CORPORATE LEADERSHIP IN SUSTAINABILITY* Dimension

| | CSR & CORPORATE LEADERSHIP IN SUSTAINABILITY | | | | |
	Sum of Squares	df	Mean Square	F	p-value
Between Groups	14.243	2	7.122	0.610	0.544
Within Groups	2803.831	240	11.683		
Total	2818.074	242			

TABLE 11.8
One-Way ANOVA Result Between Type of Outlets and OVERALL *ADOPTION OF GREEN LOGISTICS* Dimension

| | OVERALL ADOPTION OF GREEN LOGISTICS | | | | |
	Sum of Squares	df	Mean Square	F	p-value
Between Groups	32.485	2	16.243	1.187	0.322
Within Groups	3283.831	240	13.682		
Total	3316.316	242			

It is seen from one-way ANOVA result (Table 11.6), that there is no statistically significant difference in overall mean score of *ENVIRONMENTAL DEGRADATION* [$F_{(2,240)}$ = 0.237, p = 0.855, p > 0.05] dimension among three types of Retail Outlets. F value is 0.237 with degrees of freedom of 2 and 240, and the p-value is greater than 0.05 at 5% significance, strongly indicating no difference between the opinions among retailers dealing with different categories of products with regard to the importance of Environmental Degradation as regards the sustained adoption of GL. Hence, we accept the null hypothesis. In other words, mean rating score on degree of agreement with regard to *ENVIRONMENTAL DEGRADATION* dimension does not differ significantly between respondents working exclusively in *Apparel Retail* outlets and respondents working in *General & Grocery* and respondents working in *Consumer Electronic Goods* Retail Outlets.

It is seen from one-way ANOVA result (Table 11.7) that there is no statistically significant difference in overall mean score of CSR & *CORPORATE LEADERSHIP IN SUSTAINABILITY* [$F_{(2,240)}$ = 0.610, p = 0.544, p > 0.05] dimension among three types of Retail Outlets. F value is 0.610 with degrees of freedom of 2 and 240. The p-value is greater than 0.05 at 5% significance, strongly indicating no difference between the opinions among retailers dealing with different categories of products with regard to the importance of car and Corporate Leadership in Sustainability as regards the sustained adoption of GL. Hence, we accept the null hypothesis. In other words, mean rating score on degree of agreement with regard to CSR & *CORPORATE LEADERSHIP IN SUSTAINABILITY* dimension does not differ significantly between respondents working exclusively in *Apparel Retail* outlets and resplondents working in *General & Grocery* and respondents working in *Consumer Electronic Goods* Retail Outlets

There is no significant (statistically) difference (Table 11.8) in overall mean score of *Herding Effect* [$F_{(2,240)}$ = 1.187, p = 0.322, p > 0.05] dimension among three types of outlet customers. Hence, we accept null hypothesis (H_{01}) and reject alternative hypothesis (H_{11}).

11.5.6 MEASURES TO ACHIEVE CIRCULAR ECONOMY THROUGH GREEN LOGISTICS:

a. GL can be considered as one of the design solution for the logistics services. GL solutions for the retail sector not only ensure the availability of right product at the right place and at the right time but also reduces the impact of logistics operation on the environment. The design solutions for logistics operation as per CE should not only achieve the timely delivery of products but also consider the operational impact on the environment. Through route optimization the energy requirements for logistics services are optimized leading to achieving CE through reduction of consumption of energy/fuel. All logistics service design should take into account the concepts of CE and offering GL solution is a step towards this.

b. The concept of CE which stipulates preserve, extend and reduce waste can be ensured through GL solution for all logistics requirement of retail organization. Timely transportation and right flow of goods ensure that the product can be preserved and its life is extended for timely consumption.

Otherwise, shelf life of the products expire and the product cannot be consumed leading to lot of waste.

c. Retail operation require lot of collaborative interaction between various stake holders such as suppliers, manufacturers, logistics service providers, distributors, etc. in order to create value to the customers. These collaborations among various stakeholders aim to provide the best product to the consumer at the right time with the optimum use of the required resources and creating value to the customer. This is one the main objective of CE for which GL play a major role towards achieving the closed loop of CE.

d. For effective coordination in order to achieve timely delivery and optimization of inventory at each of the point along the supply chain, information flow across the entire supply chain is a pre-requisite. Towards this, the digital technology is playing a key role. Whether it is Retail or e-tail business, the new business model which require efficient, resilient and responding supply chain to be in place is only facilitated by the digital technology which undergo constant upgradation and innovation adopting newer and latest technological advances. Latest technologies such as IoT, Machine Learning, Artificial Intelligence, Blockchain, etc. are being applied for effective supply chain management. The Industry 4:0 revolution is also creeping into the supply chain management in its own pace for establishing the closed loop and achieving CE.

e. Reducing waste and optimizing the inventory level are one of the important factors for ensuring business sustainability of any retail unit and in this regard the role of logistics in reaching product at the right time, right place and in right condition cannot be re-emphasized. Towards this, the GL solution offers the necessary support for the retail business units to achieve their business goals as well as keeping the environment and ecology clean and preserved.

11.6 CONCLUSION

Government Policy & Regulation has a major influence on the adoption of GL and it can be considered as one of the main drivers for the adoption of GL by the organized retail sector. From the Indian Logistics Services context, the respondents opine that the Government should ensure that uniform norms are applicable for all types of transport, which are used for the purpose of logistics. The respondents fully asserted that if GL are not adopted, eventually it would lead to Environmental Degradation as the Logistics operation adversely affects the environment with release of GHG emissions to the environment.

Of-late, corporate entities have been entrusted mandatorily to engage themselves with activities related to corporate social responsibility. As sustainability has become one of the leading issue being raised at many platforms and industry forums, the industry is positively responding to this call and few corporates have taken a lead in this regard. Hence, the corporate entities are assuming leadership in sustainability and driving such programs as part of their corporate social responsibility. Hence, the decision by the corporate management to assume leadership in sustainability and engage in CSR activities will lead to better life and livable environment and eco system for the human existence. Thus, the Government Policy and Regulations, Environmental Degradation and the organizations assuming the Leadership in sustainability and driving the sustainability through CSR activities are the main driving forces for the sustained adoption of GL by Organized Retail Sector.

The adoption of GL by organized retail sector enable achievement of CE as GL aims to reduce GHG emission thereby reducing the adverse impact of Logistics operation on the environment. For Retail organization, flow of goods of various types are essential to run the business. Logistics services ensure the flow of goods into the retail at the time when it is most needed. The entire inventory management at any retail store is through proper coordination between the suppliers and the logistic service providers. Number of trips per day or per week is determined by the demand for particular item and timely replenishment are essential for ensuring high customer service without stock-out situations in the store. Thus Logistics plays a very crucial role for the sustainability of retail operation and GL ensures that the environmental and ecological aspects are not neglected. For the Logistics sector, adaptation of GL is a pre-requisite for achieving the CE which optimizes fuel/energy consumption and reduce the impact of logistics operation on the environment.

CONFLICT OF INTEREST

The authors declare that there is no conflict of interest for this publication.

ACKNOWLEDGEMENT

The authors would like to thank anonymous reviewers for their constructive comments to improve the quality of this paper.

REFERENCES

Aron, L. J. (1994, April). Proctor & gamble: Logistics strategy: Simplify and standardize. *Inbound Logistics*, April Issue, 24–29.

Aronsson, H., & Huge Brodin, M. (2006). The environmental impact of changing logistics structures. *The International Journal of Logistics Management*, *17*(3), 394–415.

Azapagic, A., & Clift, R. (1999). Life cycle assessment and multi objective optimization. *Journal of Cleaner Production*, *7*, 135–143.

Bacallan, J. J. (2000). Greening the supply chain. *Journal of Business and Environment*, *6*(5), 11–12.

Beamon, B. M. (1999). Designing the green supply chain. *Journal of Logistics Information Management*, *12*(4), 332–342.

Boge, S. (1994). The well-travelled yogurt pot: Lessons for new freight transport policies and regional production. *World Transport Problems and Practice*, *1*(1), 7–11.

Caputo, M., & Mininno, V. (1996). Internal, vertical and horizontal logistics integration in Italian grocery distribution. *International Journal of Physical Distribution and Logistics Management*, *26*(9), 64–90.

Corbett, C. J., & Kleindorfer, P. R. (2003). Environmental management and operations management: Introduction to third special issue. *Production and Operations Management*, *12*(3), 287–289.

Dakshina Murthy, R. A., & James, L. (2014). Study on awareness of green logistics in organized retail sector in Bangalore. *International Journal of Economics and Management
Strategy*, *4*(1), 37–44.

El-Berishy, N., *et al.* (2013, September). The Interrelation between Sustainability and Green
Logistics. *Proceedings of 6th International Federation of Automatic Control, (IFAC)
conference on Management and Control of Production & Logistics*, Vol. 46(24).
527–531.

Field, A. (2009). *Discovering statistics using SPSS*. 3rd Edition, London: Sage Publications.

Fonseca, L., Domingues, J., Periera, M., Martins, F., & Zimon, D. (2018). Assessment of circular economy within Portuguese organizations. *Sustainability*, *10*, 2521.

Fornell, C., & Larcker, D. F. (1981). Evaluating structural equation models with unobservable
variables and measurement error. *Journal of Marketing Research*, *18*(1), 39–50.

Geng, Y., Zhu, Q., Doberstein, B., & Fujita, T. (2008). Implementing China's circular
economy concept at the regional level: A review of progress in Dalian, China. *Waste
Management*, *29*(2), 996–1002. DOI:10.1016/j.wasman.2008.06.036

George, D., & Mallery, P. (2010). *SPSS for windows step by step: A simple guide and reference 17.0. Update*. 10th Edition, Boston: Pearson.

Guihua, W. (2013). Study on mutual influence and synchronized development of circular
economy and circular logistics. *Logistics Technology*. Vol. 5.

Hair, *et al.* (1995). *Multivariate data analysis*, 3rd Edition, New York: MacMillan.

Hair, *et al.* (2011). The use of partial least squares structural equation modeling in strategic management research: A review of past practices and recommendations for future
applications. *In Long Rand Planning*, *45*(5–6), 320–340.

Islam D. M. Z. *et al.* (2012). Logistics & supply chain management. *Research in Transportation
Economics*, *41*(1), 1–14.

Kumar, A., Abdul, M., Zahidur, R. L., & Angappa, G. (2019). Evaluating sustainable drivers
for social responsibility in the context of ready-made garments supply chain. *Journal
of cleaner production*. *https//doi.org/10.1016/j.jclepro.2019-119231.*

Kumar, A., Syed Abdulla, R. K., Md. Abdul, M., & Jose, A. G. (2020, March). Behavioral
factors on the adoption of supply chain practices. *Resources, Conservation &
Recycling*.

Kumar, N., Saxena, S., & Agrawal, R. (2012). Supply chain management: Road ahead with a
literature review based analysis. *Journal of Supply Chain Management Systems*, *1*(4),
37–57.

Kumar, S. *et al.* (2012). A case study from Indian electrical and electronic industry.
International Journal of Software Computing and Engineering, *1*(6), 275–281

Matthias, K. (2016). To green or not to green: A political, economic and social analysis for the
past failure of green logistics. *Sustainability*, *8*(5), 441.

Mckinnon, A. C., & Ge, Y. (2004). Use of a synchronized vehicle audit to determine opportunity for improving transportation efficiency in Supply Chain. *International Journal
of Logistics: Research and Applications*, *7*(3), 219–238.

Mckinnon, A., Edwards, J., & Cullinan, S. (2011). Comparative carbon auditing of conventional and online retail supply chain: A review of methodological issues. *International
Journal of Supply Chain Management*, *16*(1), 57–63.

Millighan, B. (2000). Transportation holds up its end of Just-In-Time bargain. *Purchasing*,
129(4), 68–75.

Murphy, P. R., Poist, R. F., & Braunschweig, C. D. (1994). Management of environmental
issues in logistics: Current status and future potential. *Transportation Journal*, *34*(1),
48–56.

Murphy, P. R., Poist, R. F., & Braunschweig, C. D. (1995). Role and relevance of logistics
to corporate environmentalism: An empirical assessment. *International Journal of
Physical and Distribution Management*, *25*(2), 5–19.

Murphy, P. R., Poist, R. F., & Braunschweig, C. D. (1996). Green logistics: Comparative views of environmental progressives, moderates, and conservatives. *Journal of Business Logistics, 17*(1), 191–211.

Murphy, P. R., & Poist, R. F. (2000). Green logistics strategies: An analysis of usage patterns. *Transportation Journal, 40*(2), 5–16.

Murphy, P. R., & Daley, J. M. (2001). Profiling international freight forwarders: An update. *International Journal of Physical Distribution & Logistics Management, 31*(3), 152–168.

Murphy, P. R., & Poist, R. F. (2003). Green perspectives and practices: A "comparative logistics" study. *Supply Chain Management: An International Journal, 8*(2), 122–131.

Murray, A., Skene, K., & Haynes, K. (2017). The circular economy: An interdisciplinary exploration of the concept and application in global context. *Journal of Business Ethics, 140*, 369–380.

Naustdalslid, J. (2014). Circular economy in China – The environmental dimensions of the harmonious society. *International Journal of Sustainability Development of World Ecology, 21*, 303–313.

Ndubisi, N. O. (2006). Effect of gender on customer loyalty: A relationship marketing approach. *Marketing Intelligence & Planning, 24*(1), 48–61.

Oksana, S., & Agniezka, O. (2019). Green logistics and circular economy. *Transportation Research Procedia, 39*, 471–479.

Park, J., Sarkis, J., & Wu, Z. (2010). Creating integrated business and environmental value within the context of China's circular economy. *Ecological Modernization, 18*, 1494–1501.

Potter, A., Mason, R., & Lalwani, C. (2007). Analysis of factory gate pricing in the UK grocery supply chain. *International Journal of Retail & Distribution Management, 35*, 821–834.

Rondinelli, A. D., & Berry, M. A. (2000). Environmental citizenship in multinational corporations: Social responsibility and sustainable development. *European Management Journal, 18*(1), 70–84.

Roberta, D.A, Howard, M. & Joe, M. (2018). SCM and the Circular Economy: towards the circular supply chain. *Journal of Production Planning & Control, The Management of Operations, 29*(6), 425–437.

Rong, Z. (2014). Study on the sustainable development of logistics for circular economy. *Proceedings of the International Conference on Logistics, Engineering, Management ad Computer Science (LEMCS -2014)*. Atlantis. https://doi.org/10.2991/lemcs-14.2014.10

Rothenberg, S., Pil, F. K., & Maxwell, J. (2001). Lean, green and the quest for superior environment performance. *Production & Operations Management, Fall, 10*(3), 228.

Sarkis, J., Meade, L. M., & Talluri, S. (2004). E-Logistics and the natural environment. *Supply Chain Management, 9*(4), 303–312.

Sbihi, A., & Eglese, R. W. (2009). Combinatorial optimization and green logistics. *Annals of Operations Research, 175*(1), 159–175.

Skjoett-Larsen, T. (2000). European logistics beyond 2000. *International Journal of Physical Distribution and Logistics Management, 30*(5), 377–387.

Smithers, R. (2007, September). Supermarket home-delivery services promotes its green credentials. *Guardian dated 12-09-2007*, 12.

Srivastava, S. K. (2007). Green supply-chain management: A state-of-the-art literature review. *International Journal of Management Reviews, 9*(1), 53–80.

Svensson, G. (2007). Aspects of sustainable supply chain management (SSCM): Conceptual framework and empirical examples. *Supply Chain Management, an International Journal, 12*(4), 262–266.

Stahel, W. R. (2016). The circular economy. *Nature, 531*, 435–438.

Sunil, L., Gunjan, Y., Kumar, A., Anthony, A., Sachin, K. M., & Dixit, G. (2020, March). Study on the key enablers of Industry 4:0 practices implementation using ISM – Fuzzy MICMAC approach. Proceedings of the International conference on Industrial Engineering and Operations Management, UAE, Dubai, March 10-12, 2020, pp. 241–251

Su-Yol Lee & Klassen, R.D. (2008). Drivers and enablers that foster environmental management capabilities in small and medium sized suppliers in supply chains. *Journal of Production & Operations Management* , *17*(6), 573–586.

Szymankiewicz, J. (1993). Going green: The logistics dilemma. *Journal of Logistics Information Management*, 6(3), 36–43.

Whitelegg, J. (1993). Time pollution. *The Ecologist*, *23*(4), 131.

Wu, H. J., & Dunn, S. C. (1995). Environmentally responsive logistics systems. *International Journal of Physical Distribution and Logistics Management*, 25(2), 20–38.

Yuan, Y., & Xi-long, Y. (2009). Logistics operation based on circular economy. *Environmental Protection of Xinjiang.*

Zheng, L., & Zheng, J. (2010). Research on green logistics system based on circular economy. *Asian Social Sciences*, *6*(11), 116–119.

Zink, T., & Geyer, R. (2017). Circular economy rebound. *Journal of Industrial Ecology*, *21*, 593–602.

Zou, X., Zhang, M., & Chen, H. (2015). Analysis on the connotations of green logistics. Published by *Atlantis Press for International Symphosium on Social Science.* 130–136

12 Framework for Sustainable Food Systems
Holistic Mitigation and Adaptation

Sarika Yadav
Department of FBM & ED, National Institute
of Food Technology Entrepreneurship and
Management, Sonepat, Haryana, India

Rahul S. Mor
Department of Food Engineering, National
Institute of Food Technology Entrepreneurship
and Management, Sonepat, Haryana, India

Simon Peter Nadeem
Centre for Supply Chain Improvement,
University of Derby, Derby, UK

CONTENTS

12.1 INTRODUCTION

Food, being the basic need and metabolic fuel for the body, needs to have a sustainable supply chain. While this concept has been studied and applied in food chains in certain countries across the globe, it is still relatively new in many developing nations. Here, increasing populations challenge food security by the issues they create for ensuring the availability, palatabilty and, importantly, distribution of food. Agri-food marketplaces are still incompetent and the food system is powered by fluctuations in protocols and consumption patterns, economic rivalry among processors and supermarkets as well as the pressure from international competitors (Prakash, 2019). A sustainable supply chain provides an efficient and effective distribution channel. The distribution of food accounts for a large amount of energy consumption and has its effect on the environment as well (Mor *et al.*, 2019; 2018a). The average distance in the United States alone is about 1,500 miles and in some cases, food can travel up to 3,100 miles (Pirog *et al.*, 2001; Pimentel *et al.*, 2008). This step is by far means the feeblest link in food chains because of the excessive usage of fossil fuels, principally diesel that causes the emission of gases that impact the environment at large. Further, the more massive and slower a vehicle is, the lesser is the energy it consumes and vice versa (Morawicki, 2011). Therefore, a sustainable solution is the need of the hour for the distribution of fresh fruits and vegetables since these cannot be transported by slow-moving carriers due to their perishability nature.

The traditional supply chain model of fresh fruits and vegetables in India includes a large number of stakeholders in the form of intermediaries and other agents who would lose their jobs and source of income if an alternate model, that eliminates these intermediaries, comes into existence and is popularized. The changes that happen in the food system have an impact on the supply chain factors and the sustainable food systems in general (Prakash, 2019). Also, most consumers still prefer purchasing fresh produce in person since they get to feel and visualize what they are buying and this gives them a sense of satisfaction concerning the quality of the produce. The fresh fruits and vegetable sector in India is still mostly unorganized. Hence, a lot of coordination is required to integrate small vendors and other unorganized players within a model. A few big names that belong to the organized sector and these companies have a relatively efficient supply chain and hence, prime focus has to be on small scale retail outlets and street vendors.

12.2   SUSTAINABLE FOOD DISTRIBUTION

India, being the leading producer of fruits and vegetables globally, has the capability of becoming the food basket of the world if the magnitude of post-harvest losses is kept under control and minimized. The enhanced consumer awareness with development, trade globalization and agro-industrialization recently has led to an exponential growth of sustainable food supply chains (Darbari *et al.*, 2018).An important consideration to develop a sustainable solution for the distribution of fresh fruits and vegetables is the proximity. The more proximate a market is from the consumer, the lesser will be the transportation costs, inventory costs, and so on. Moreover, if transportation is minimized, the pollution caused to the environment will also be reduced. Lesser inventory costs will ensure the minimization of the use of electricity for the storage of fresh fruits and vegetables in controlled atmosphere storage systems. However, for seasonal produce, controlled atmosphere storage is an excellent option to preserve the produce in its natural state for long periods. High electricity costs are the most important limiting factor here. If local vendors are integrated within a planned model, sustainability for the sale of fresh fruits and vegetables can be ensured to a certain extent. In the Indian scenario, local vendors play an important role in food sales and distribution. This is because the food industry is dominated by the unorganized sector which accounts for 85%. Therefore, involving these small players would be the best direction in which to move ahead. These local street vendors usually stack the fresh produce in push-carts and sell their products in different residential colonies. The fresh produce is under the direct sun which will increase its rate of deterioration. Moreover, it is common that more than one vendor will cover the same area. This is irrational due to the wastage of time and deterioration of the produce. Therefore, a feasible solution can be the integration of these local vendors by forging a common understanding amongst them and also by establishing a better customer relationship with their buyers. Another aspect that can be brought to the fore here is the incorporation of sustainability within these existing cold chains or refrigerated systems. With the advancement of technology in the field of mechanical engineering, systems have been developed that can utilize the heat liberated from truck engines through absorption cycles. One such prototype was developed by Koehler who engineered this system for a truck trailer that ran on exhausted gases of an engine that consumed diesel. This developed system was capable of producing a cooling capacity of nearly 5 kW with a coefficient of performance of 0.27 (Tassou *et al.*, 2009). Another sustainable solution aims at harnessing solar energy by the use of solar photovoltaic cells that can be installed on the roof of transport vehicles involved in a cold chain. Such work was autonomously carried out in the United Kingdom's University of Southampton and the United States' Sandia National Laboratories (Bahaj, 2000; Bergeron, 2001).

The first aspect of this integrated system can be implemented by allocating a certain said area to the vendors for the sale and distribution of their produce. Measures can also be undertaken to ensure the availability of variants of products among the vendors and exchange of produce that is unavailable with a vendor but available with another vendor by mutual understanding.

The second aspect of the system borrows a broken-down concept of e-commerce. E-commerce is the sale of commodities via placing orders online and delivery to the desired location. The concept of e-commerce has illustrated the delivery of quality products, reduction in lead times, and reduction of inventory levels at the retailer end. The use of this fact can be used to develop a system wherein customers can order for particular fresh fruits and vegetables of their choice in a defined quantity. This will reduce the wastage of produce due to deterioration when exposed to direct sunlight and pollution during the sale and distribution. Further, vendors can also pack the produce for advance-orders in plastic bags and the plastic bags can be reused after collecting during the next delivery. This will ensure environmental sustainability as well by adhering to the concept of reduce, reuse and recycle. These local vendors can also increase profits by carrying out minimal processing activities such as cleaning, washing, grading, sorting, cutting the fresh fruits and vegetables and packing them into containers for ease of use for customers who are pressed for time due to hectic work schedules (Srivastava *et al.*, 2011).

12.3 SUSTAINABLE FOOD SYSTEMS: A WAY FORWARD

Feeding a population as large as India's will require innovation as well as adroitness. The aim shall be the 'sustainable way of producing sufficient food holistically with limited resources'. The major role to play in this scenario would be that of the policymakers along with guidance from international agencies. India from the advent of its green revolution is a country obsessed with grain production as a national measure of food security. This means has proved inconsistent leading to crippling price inflation and nutrition scarcities. To survive the competition and accelerated sustainable performance, food enterprises are executing proactive approaches. Though, there exist numerous barriers to the effective implementation of the sustainable food supply chain in India (Darbari *et al.*, 2018). There is a need to assent a demand-driven approach for accelerated growth and food sustainability. The environmental and economical aspect has to be kept in purview while framing models and policies. Food industries must transform to adopt a change in climate as this affects the raw material supply, to meet the demands, and to reduce the emission intensities per output (Mor *et al.*, 2018b). Strategies have to be fringed that are climate responsive and can achieve triple wins of the 'Development, Adaptation and Mitigation'. For achieving enhanced sustainability throughout the food supply chain system, it is imperative to work on common objectives in the three dimensions, i.e. economic, environmental and social (Perales *et al.*, 2019). Food industries are a paramount threat to environmental disturbances due to their high effluent emissions. Dependence of food industries on already limited resources also raises the vulnerability of agri-dependent communities. An effective strategy shall be to step up resilience by preparing the communities to deal with resource scarcity and extreme events (Swain, 2014). Mitigation in agri-based industries would require rectifying efficiency in resource consumption such hat sudden stress, climate change and extreme natural events can be decreased. In this line, adaptation and mitigation have a similar objective, looking for sustainability in food production and consumption. This goal is achievable in terms of a food system that encourages micro-entrepreneurship. With the help of governmental subsidies and policies and educated youth empowerment towards food systems, a network of the self-sustainable system could be formed that add zero percent to ecological footprint.

12.4 MICROENTERPRISE MATRIX FOR HOLISTIC MITIGATION AND ADAPTATION

The concept 'microenterprise matrix for holistic mitigation and adaptation' refers to a small business that is efficient in adding value to a nation's economy by generating jobs, augmenting income, supporting purchasing power, and adding business convenience. A matrix of micro-food business when chained together with an array of local producers as well as processors backed by government policies, extension services and subsidiaries, can prove to solve the food tribulations of a bucolic country like India. Moreno-Luzon *et al.* (2019) proposed a theoretic model to link supplier integration as a dynamic competence to ordinary routines. The terms to support the model are foothold to a system that is simple and yet can prove to play a hefty role in lacerating the existent ecological footprint and sustaining a food availability to current and future needs. The system is self-sufficient, workable, flexible to changes, and a major reducer for the carbon footprint. The concept of microenterprise arises from the foremost step in carbon emission reduction to which the transportation of goods adds a large number. Only if the food supply chain is managed well, can greenhouse emissions be reduced by an enormous amount. The second reason for choosing this concept for the Indian food system is that most of the population depends on agriculture food their livelihood. The microenterprise matrix not only eliminates the middleman but also gives a better price for the goods. The model has a retail store as a protagonist and the store would depend on the local good providers as well as giant enterprises of food industries for bulk goods like grains and processed food that is not available locally. For the maximum number of goods, it will depend on the local processing houses, which would maintain a standard according to Indian food regulations of FSSAI (Food Safety and Standard Authority of India). The opening of the retail store shall subsidize by Indian Govt., especially in rural areas. The diameter of coverage of one retail store shall not exceed 5 km depending on the consumption and population of the area. This model is supported by the recent development in a retail scenario in India as the government has allowed FDI (Foreign Direct Investment) in the sector.

The model constitutes the elimination of middlemen who would earlier take a major margin of profit by selling raw Agri commodities. The street and cart vendors are the registered food business operators who can get commodities from a retail store at wholesale prices to sell further to consumers. On the part of consumers, they may have a facility of doorstep service through e-commerce of the local retails store for which they would not have to go to the retail store to buy the daily goods. This saves fuel of a considerable number of consumers and thus cuts the carbon emission. The responsibility on the consumer's part is to efficiently manage the waste they produce as a result of the consumption of food material. A community biogas plant can be set up to use the bio-waste and turn into manure and biogas for use as a fuel. Secondly, the recyclable material like glass, tin, and paper shall be separated and recycled as well. Dobon *et al.* (2011) highlighted the environmental evaluation of a flexible best-before-date communicative device on a packaging consumer unit and its inferences on reducing eco-friendly impacts for fresh products. The big enterprises can be the supplier of grains and pulses and processed foods that are not locally available but is in demand. The large bulk of goods can be sent in fewer frequencies in all the local retail stores in the vicinity.

12.5   CONCLUSION

The microenterprise food sustainable model is a promise to provide job opportunities to rural youth, women empowerment and mitigation in ecological footprint. To adapt this model as a success, the government has an important role to play in terms of an increasing share of FDI, policies and subsidiaries. The model undertakes inclusive, faster and sustainable growth of food security with alleviating existing food problems. Consumers have a role to play in terms of waste management and utilization of existing water resources diligently. The four proposed problems of irrigation and fertilizers, water management, livestock and traditional measures form the basis of the above-stated model. The model is flexible and can constitute accumulation and reinforcement of other food system strategies along with immediate sustainable actions. The model is simple yet effective in sowing the seeds for a sustainable distribution.

However, the training and education highlighting these sustainable solutions are the need of the hour and is highly essential to ensure food security. Sufficient levels of awareness need to be created among small scale vendors about the management of post-harvest losses and various measures to overcome them. The initiatives by the government authorities in terms of subsidies and other financial assistance should be conveyed to the common man through effective means. Also, various skill development programs must be encouraged by highlighting their need and manner in which it will benefit the participants.

REFERENCES

Bahaj, A. S. (2000, September). Photovoltaic power for refrigeration of transported perishable goods. In Conference Record of the 28th IEEE Photovoltaic Specialists Conference-2000 (Cat. No. 00CH37036) (pp. 1563–1566). IEEE.

Bergeron, D. (2001). Solar powered refrigeration for transport applications—A feasibility study. Final Report-SAND 2001-3753, December 2001. Sandia National Laboratories Albuquerque, New Mexico 87185 and Livermore, California 94550.

Darbari, J. D., Agarwal, V., Sharma, R., & Jha, P. C. (2018). Analysis of impediments to sustainability in the food supply chain: An interpretive structural modeling approach. In *Quality, IT and business operations* (pp. 57–68). Singapore: Springer.

Dobon, A., Cordero, P., Kreft, F., Østergaard, S. R., Robertsson, M., Smolander, M., & Hortal, M. (2011). The sustainability of communicative packaging concepts in the food supply chain. A case study: Part 1. Life cycle assessment. *The International Journal of Life Cycle Assessment, 16*(2), 168–177.

Mor, R.S., Bhardwaj, A., Singh, S., & Nema, P.K. (2019). Framework for measuring the performance of production operations in dairy industry. In Essila, J. C. (Ed.), *Managing Operations Throughout Global Supply Chains*, Hershey PA: IGI Global, https://doi.org/10.4018/978-1-5225-8157-4.ch002.

Mor, R.S., Bhardwaj, A., & Singh, S. (2018a). A structured-literature-review of the supply chain practices in dairy industry. *Journal of Operations and Supply Chain Management, 11*(1), 14–25.

Mor, R.S., Bhardwaj, A., & Singh, S. (2018b). A structured literature review of the Supply Chain practices in Food Processing Industry. *Proceedings of the 2018 Int. Conference on Industrial Engineering and Operations Management, Bandung, Indonesia*, March 6–8 (pp. 588–599).

Morawicki, R. O. (2011). *Handbook of sustainability for the food sciences.* Arkansas: John Wiley & Sons.

Moreno-Luzon, M. D., Escorcia-Caballero, J. P., & Chams-Anturi, O. (2019). The integration of the supply chain as a dynamic capability for sustainability: The case of an innovative organic company. In *Knowledge, Innovation and Sustainable Development in Organizations* (pp. 97–111). Cham: Springer.

Perales, D. P., Verdecho, M. J., & Alarcón-Valero, F. (2019, September). Enhancing the sustainability performance of agri-food supply chains by implementing Industry 4.0. In *Working conference on virtual enterprises* (pp. 496–503). Cham: Springer.

Pimentel, D., Williamson, S., Alexander, C. E., Gonzalez-Pagan, O., Kontak, C., & Mulkey, S. E. (2008). Reducing energy inputs in the US food system. *Human Ecology, 36*(4), 459–471.

Prakash, G. (2019). Exploring innovation and sustainability in the potato supply chains. In *Environmental sustainability in Asian logistics and supply chains* (pp. 97–120). Singapore: Springer.

Pirog, R. S., Pelt, T. V., Enshayan, K., & Cook, E. (2001). *Food, Fuel, and Freeways: An Iowa perspective on how far food travels, fuel usage, and greenhouse gas emissions.* Leopold Center Pubs and Papers. 3. http://lib.dr.iastate.edu/leopold_pubspapers/3

Srivastava, R. C., Ambast, S. K., & Ahmed, S. Z. (2011). *Vision 2030.*, Port Blair, A & N Islands: Central Agricultural Research Institute.

Swain, A. K. (2014). Can India Reform Its Agriculture? *The Development,* Retrieved from: https://manage.thediplomat.com/2014/06/can-india-reform-its-agriculture, July 15, 2020.

Tassou, S. A., De-Lille, G., & Ge, Y. T. (2009). Food transport refrigeration– Approaches to reduce energy consumption and environmental impacts of road transport. *Applied Thermal Engineering, 29*(8–9), 1467–1477.

13 Supply Chain Network Design Models for a Circular Economy
A Review and a Case Study Assessment

Sreejith Balasubramanian, Mahshad Gharehdash, and Mahnoush Gharehdash
Middlesex University Dubai, Dubai, UAE

Vinaya Shukla
Middlesex University, London, UK

Arvind Upadhyay
Brighton Business School, University of Brighton, Brighton, UK

CONTENTS

13.1 INTRODUCTION

Globalization provides international firms' with a host of benefits such as access to new markets and reduced costs of production and taxes (through manufacturing across borders) (Varzandeh *et al.*, 2016). However, this requires management/ leveraging of global supply chain networks (Garcia & You, 2015) that are facing pressure from competitors and customers. These networks are, therefore, constantly being reconfigured so that they continue to be competitive. However, this is making them increasingly complex to manage (Hammami *et al.*, 2008; Jaehne *et al.*, 2009; Mokhtar *et al.*, 2019) and also increasing the risks of disruptions (for multinational firms). For example, the current COVID-19 pandemic has caused large scale travel and trade restrictions with associated disruptions and consequences for global supply chains across most sectors. Similar is the case for international political instability related (negative) implications for global supply chains. For example, Brexit has caused disruptions in the global supply chains of many firms (Lockett *et al.*, 2019), and Apple Inc. is considering moving its manufacturing from China to India due to the US–China Trade war (Vaitheesvaran, 2019). Resilient supply chains with the ability to handle such disruptions have a significant strategic advantage over the others that don't.

The other critical aspect is environmental sustainability. Managers are under increasing pressure to reduce their supply chain's adverse environmental impacts. This is not surprising given that environmental pollution and global climate change have emerged as one of the significant challenges of the 21st century. Thus, industries around the world are looking at options to meet the market demand in a more environmentally responsible way (Habib *et al.*, 2020).

The rubber sector is one of the key ones from an environmental impact perspective. Rubber-based industries have witnessed significant worldwide growth in recent times (Chanchaichujit *et al.*, 2020), and that is also expected to continue into the future; annual growth rate of around 5% is projected for the next ten years to reach a market size of USD 45 billion globally by 2027 (Kenneth Research, 2019). This is largely due to the growing demand from the tire industry. Rubber production is considered to be energy-intensive and environmentally polluting (Jawjit *et al.*, 2015), though there have been few efforts to tackle its negative environmental impacts (Chanchaichujit *et al.*, 2020).

Therefore, a great deal of emphasis is now placed on the design/redesign of global supply chains (dispersed geographical elements that are highly coordinated with each other) that takes into account environmental sustainability, cost, responsiveness and supply chain risk considerations, together with the short term and long term organisational objectives. Figure 13.1 shows the different supply chain network design/redesign aspects that need to be simultaneously considered.

This increased relevance given to supply chain design/redesign is reflected in the growing academic interest in this area, especially in the use of mathematical models for supply chain design/redesign. Yet, different studies have used different approaches, tools and techniques for supply chain design/redesign, and that too with very specific objectives, resulting in the knowledge being fragmented and scattered across the literature. A holistic understanding of global chain modeling strategies

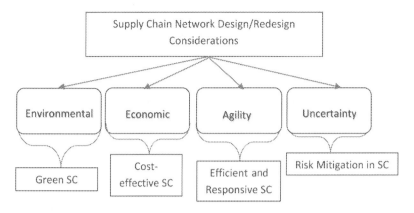

FIGURE 13.1   Supply chain network design considerations.

is therefore lacking, and which requires a comprehensive review to understand the current state of knowledge on this subject. This forms the focus of this study whose objectives are:

- To conduct a review and synthesis of different mathematical modeling approaches used in supply chain management in order to comprehend the various considerations (environmental, cost, efficiency and risks) in supply chain network design/redesign
- To apply this understanding on supply chain network design/redesign to a real-world case study of a multinational tire manufacturing firm
- To provide implications for research and practice

The rest of the study is structured as follows. The research methodology is discussed in Section 2. Section 3 presents a review of the literature on different mathematical models used in supply chain management. Its application for the supply chain redesign of the case company (a leading global tire manufacturer) is discussed in Section 4. The study concludes in Section 5, where the implications for research and practice are discussed.

13.2   RESEARCH METHODOLOGY

In line with the research objectives, the methodology adopted in this chapter consists of two parts, the literature review and the case study. Figure 13.2 shows the research methodology adopted in this study.

13.2.1   LITERATURE REVIEW

The review of the literature was undertaken using the Web of Science database. The keywords used for the search included 'Global Supply Chain Network Design' and 'Global Supply Chain Network Redesign'. While this returned more than

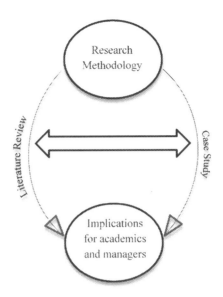

FIGURE 13.2   Research methodology used in study.

1,000 studies, these were narrowed down by abstract-based screening; only studies that had used mathematical modeling and had a primary focus on one of the research objectives were retained, while the others were excluded.

13.2.2   CASE STUDY METHODOLOGY

The case study considered is the regional headquarters of a global tire manufacturing firm. It is located in Dubai, UAE, and is responsible for meeting the demands of 52 countries in the Middle East and North Africa (MENA) region.

An initial two-hour meeting with the CEO and senior supply chain managers was conducted to understand the significant issues in their supply chain network. This was followed by meetings with other senior executives where internal company data was also obtained. Additionally, four semi-structured interviews with mid-level executives from the different departments were conducted to comprehend their operational supply chain issues. Further clarifications were obtained via emails and phone calls.

13.3   LITERATURE REVIEW FINDINGS

13.3.1   CONSIDERATIONS FOR SUPPLY CHAIN NETWORK DESIGN/REDESIGN

13.3.1.1   Environmental/Green Supply Chain Models

Governments, organizations and business managers are facing increasing pressure to minimize the environmental impacts of supply chains. This is because the related issues of environmental pollution, climate change and resource depletion have become one of the greatest challenges of the 21st century (IPCC, 2007).

The total global greenhouse gas (GHG) emissions, the main driver of climate change, amounted to approximately 52.7 gigatonnes of carbon dioxide equivalent (GtCO2e) in 2014, the highest level reported since the pre-industrial levels (UNEP-EGR, 2014). Also, the increase in the annual rate of GHG emissions during the period 2000–2010 was faster (2.2%) than during the period 1970–2000 (1.3%) (UNEP-EGR, 2016). The effects of these emissions, mainly in the form of global warming and rising sea levels, are clearly evident: 2015 was the hottest year ever recorded and ten of the warmest years on record have occurred since 2000 (UNEP-EGR, 2016); the rate of rising sea levels has accelerated in recent years (EPA, 2017). The significant push for economic development and industrialization is also accelerating the depletion of natural resources. At current rates of use, the world will soon run out of many vital resources, including renewable resources. For example, assessment by the US Energy Information Administration (EIA) shows that fossil fuels could be entirely depleted in the next 25 years (EIA-IEO, 2013). From a rubber industry perspective, too, there is increasing pressure on the industry from global buyers, especially in developed countries, to reduce the negative environmental impacts (Krungsri Report, 2019). However, this needs to be done while still meeting the increasing global demand for rubber products and sustaining its economic contribution.

Because of this, incorporating environmental concerns into supply chain management, or green supply chain management (GSCM), has seen significant interest among academics and practitioners. GSCM addresses environmental issues dispersed across the different stages of the supply chain, i.e. from design through to end-of-life leading to a circular economy. Trade-offs are usually involved, and generally among incompatible objectives; optimization models for GSCM-related decision making are therefore common (Ansari & Kant, 2017). These models are based on mathematical procedures that strive to find optimum solutions under a given set of assumptions, constraints and data (Coyle *et al.*, 2004). While different mathematical programming procedures/techniques such as linear programming, mixed-integer programming and non-linear programming (Ansari & Kant, 2017; Srivastava, 2007) are used, single and multi-objective linear programming techniques are the most popular (Ansari & Kant, 2017).

Chanchaichujit *et al.* (2016), for example, used the single objective linear programming model to find the association between the quantity of rubber product flow between supply chain entities and the transportation mode and route, to minimize total GHG emissions. The authors considered GHG emissions and costs as two single objective functions and found the relationship between GHG emissions and costs to be in conflict with each other. Multi-objective optimization has also been considered by different researchers; this involves minimizing total costs or maximizing total profits while simultaneously minimizing environmental impacts (Kim *et al.*, 2009; Wang *et al.*, 2011). In these studies, the total costs are usually the summation of supply chain activity costs such as production, inventory, and transportation (You & Wang, 2011), while total profits are generally the net profits (Hugo & Pistikopoulos, 2005). For environmental objectives, different measures are used including CO2 emissions (Kim *et al.*, 2009; Wang *et al.*, 2011), GHG emissions (You & Wang, 2011), energy consumption (Winebrake *et al.*, 2008) and Global Warming Potential (Buddadee *et al.*, 2008).

13.3.1.2 Closed-Loop Supply Chain Models

A related aspect to environmental supply chains is the closed-loop supply chains. Closed-loop supply chain design refers to the integration of the forward and reverse supply chain designs (Amin *et al.*, 2017; Pedram *et al.*, 2017). Reverse supply chains involve activities that return used goods to the manufacturers, for reuse/refurbishing/ remanufacturing. In the case of the tire industry, this involves the tire being returned, retreaded and made available in the market again (Amin *et al.*, 2017). However, evidence from the literature shows that while the practice of tire retreading is good from an environmental standpoint, it is not preferred by manufacturers as it may reduce new product sales. Table 13.1 shows select studies in supply chain network design using closed-loop models and that have focused on the tire industry. All three studies have used multi-objective and integrated models.

Amin *et al.*'s article provide excellent insight into the tire remanufacturing process in Canada. Additionally, their framework uses scenario analysis tree to analyze sources of uncertainty (such as demand) over the multi-period decision making and considers discounted cash flow for assessment. The strategic model selects facility locations as well as those of retailers and drop-off depots; however, it does not take waste management and CO_2 emission factors into account. Furthermore, the model has been applied to a case study context that operates predominantly in one state of Canada.

The closed-loop design model of Pedram *et al.* (2017), unlike Amin *et al.*'s, considers the uncertainty of demand and introduces a formulation that solves for facility locations, including drop-off depots locations as well as product flows between subsidiaries to minimize the overall cost of transportation. The paper discusses product recovery for a tire manufacturer in Iran, and first uses mixed-integer linear programming with a fixed cost, capacity and distance between facilities. This is followed by a scenario analysis that considers three scenario expectations for demand. Nevertheless, Pedram *et al.* (2017)'s model does not account for global supply chain variables, such as duty and tax charges.

TABLE 13.1
Supply Chain Network Design Using Closed-loop Models in the Tire Industry

Author	Decision Levels		Optimization	Model	Objective
	Operational	Strategic			
Amin *et al.* (2017)	✓		✓	Mixed Integer Linear Programming and Scenario Analysis	Optimize profit of returns
Pedram *et al.* (2017)	✓		✓	Mixed Integer Linear Programming and Scenario Analysis	Optimize transportation Cost
Subulan *et al.* (2015)		✓	✓	Mixed Integer Linear Programming	Optimize profit with less environmental impact

Subulan *et al.* (2015)'s study discusses the various options of used tire recovery and disposal; it then integrates measures of environmental impact into a closed-loop supply chain logistics network design model. The framework, similar to other studies discussed in this section, uses mixed-integer linear programming to maximize profit, but in addition, it seeks to reduce the total ecological bearing of the supply chain. The model uses parameters for demand, cost (inventory, inventory hold, distribution and new facility) and capacity, then proceeds to conduct sensitivity analysis using the Taguchi Method (Robust Design). Taguchi design is a productivity method that reduces production cost, environmental impact, and seeks to meet customer demand. In this case, Subulan *et al.* (2015) used the Taguchi method to analyze the profitability of the model. The proposed framework, in addition to disregarding global factors of a supply chain, also uses many constraints, which reduces its utility for real world applications.

13.3.1.3   Cost-Efficiency Optimization Models

Cost optimization can be achieved in various ways in the supply chain; however, in most studies, this is realized by monitoring the activities of warehousing, inventory management, and order management (Croxton & Zinn, 2005). Among different frameworks, a useful one for strategic cost optimization modeling is the strategic safety stock supply chain concept developed by Graves and Willems (2000). It focuses on minimizing inventory costs and providing a high level of service for the customers in the supply chain, that is subject to demand and forecast uncertainty. The emphasis is on designing a supply chain network where each stage of the network operates with a periodic-review base-stock policy, and where demand is bounded, and service for customers, guaranteed (as shown in Figure 13.3).

Graves and Willems's model used an optimization algorithm for inventory modeling. Based on calculated assumptions, the model detects problems within the supply chain and capture the source of the problem and formulate that into deterministic optimization. The optimization algorithm minimizes the inventory cost level, meaning that the holding cost for safety stock in the supply chain decreases.

Margolis *et al.* (2018) managerial decision support supply chain network design model considers both costs, which are associated with supply chain network design decisions, such as increasing number of facilities or closing down facilities, and cost

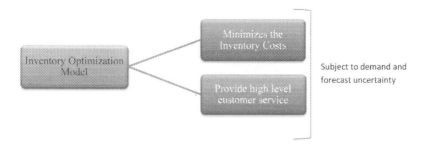

FIGURE 13.3   Inventory optimization model outcomes subject to demand and forecast uncertainty.

of operations such as assigning the production capacity and flow of products for each node through the network; its goal is to suggest optimum location and production quantity plans as well as the routing network. The authors consider weighted demand parameters to make the model less susceptible to disruptions. They also show (through examples) that their model can withstand production disruptions; nevertheless, the resultant computations are quite complex and take a long time (approximately 24 hours as per the authors for the context they considered).

Hammami and Frein (2014) analyze the previous literature on both supply chain network design and redesign; they recognize that most of it focuses on facility allocations, but which does not consider taxation, exchange rate and transfer pricing related aspects that are important in a global context. The authors acknowledge that tax rewards by some countries are the reason why many supply chain networks are redesigned by firms. Their model therefore takes a profit-maximizing approach that considers facility relocation and productions capacities, as also transfer pricing aspects. Two frameworks for transfer pricing are considered; in the first, finished products are considered where a range of prices (maximum and minimum) for the product are assumed based on similar products in the market; in the second, the focus is on semi-finished products where a Profit-Split method is assumed, where the profit is divided between the parties involved in production based on their respective contributions. Factors that impact redesign, such as facility closing cost and capacity relocation are also considered in the study.

Finally, Creazza et al. (2012) propose a supply chain network redesign optimization framework based on Pirelli Tires in Europe. Their framework uses a standard Mixed Integer Linear Programming approach to reduce cost and to reconfigure the network to realize optimum cost efficiency at the lowest cost. However, it has limitations; for example, aspects such as exchange rates and tariffs were not considered; additionally (and which is also recognized by the authors), accurate implementation of the model requires precise data, such as on transportation cost, which may not be possible to get.

Besides the individual limitations considered in each of the above models/studies, all of them have an additional limitation; they may not perform well under uncertainty, such as non-stationary demand (in place of the stationary one that is generally assumed), use of different review periods for stocks (vis-à-vis the fixed/standard one assumed), and capacity constraints that are usually ignored in models. To address these issues, a few studies have proposed cost-optimization models under uncertainty. For example, Fattahi et al. (2018) model proposes a multi-stage stochastic model that addresses the uncertainty of demand by considering different customer segments. Such a model is useful for making supply chain decisions both at strategic as well as tactical levels.

13.3.1.4 Supply Chain Risk Management Models

Global supply chains are susceptible to risks and uncertainties; this section of the paper reviews supply chain network design and redesign models that address risks and uncertainties. Goh et al. (2007) classified risks into two categories, as shown in Figure 13.4, strategic risks and operational risks. Strategic risks refer to threats of political instability (change in policies), natural disasters and climate effects that

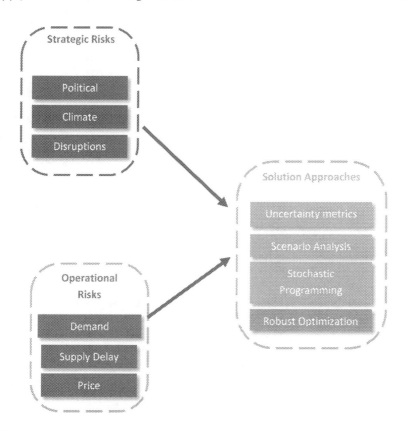

FIGURE 13.4 Levels of risks and solution approaches.

could cease production. In contrast, operational risks involve factors that could impact the operation of the supply chain, for instance, demand fluctuations, price and cost volatility. There are many approaches to solving the issue of supply chain risks; the following frameworks consider diverse categories of risks and uncertainties in their supply chain network design and redesign models.

Rahimi *et al.*'s (2019) study addressed demand uncertainty and used a two-stage stochastic programming framework to handle the fluctuation of demand and strategic risks. Stochastic programming (two or multi-stage) enables long-term scenario analysis, which helps organizations develop strategies to manage uncertainties for the long term. Here stages of design or redesign are divided into various scenarios; first, the primary decisions are made, and subsequently assessed with the dimension of uncertainty and then updated in the following stage; however, the greater the number of stages, the more is the computation time required (Garcia & You, 2015). Table 13.2 provides a summary of supply chain network design or redesign models that consider uncertainty and risks in modeling.

Rahimi *et al.*'s (2019) study considers uncertainty and risk in supply chain network design while simultaneously factoring in sustainability elements. They use a mixed-integer non-linear program that considers the carbon footprint impact of opening a

TABLE 13.2
Supply Chain Network Design Risk Models

Author	Operational Risks		Strategic Risks	Model	Objective
	Demand	Price			
Rahimi et al. (2019)	✓		✓	Two-Stage Stochastic Model	Risk Aversion Parameters
Jahani et al. (2018)	✓	✓		Multi-tier Stochastic Model	Risks due to price and demand uncertainty
Carvalho et al. (2012)			✓	Simulation Model for Resilience	Risk Minimization
Nickel et al. (2012)	✓		✓	Multi-Stage Stochastic Model	Risk Minimization
Goh et al. (2007)	✓		✓	Multi-Stage Stochastic Model	Risk Minimization

new manufacturing plant or distribution center, as well as the effect of transportation-related emissions. The model conducts a scenario analysis and assumes the worst-case scenario (total loss) in each computational stage one by one to solve the multi-objective problem. Although this study's contribution to supply chain network design is significant, the suggested model has not been applied/tested on a real-life supply chain context.

Jahani et al. (2018) propose a stochastic model that addresses risks due to price and demand uncertainty. The multi-tiered, multi-period stochastic model considers geometric Brownian motion, an exponential time-continuous model. The study then successfully uses the model to redesign the supply chain network of the Australian cement industry as a case study. However, the model falls short, as most network redesign models do; it only considers the risk and uncertainty of two dimensions, price and demand.

The objective of the framework proposed by Carvalho et al. (2012) is to incorporate a simulation-based model into the supply chain design decision-making process, and use it as a tool to boost supply chain resilience against possible disruptions caused by changes in external supply chain policies. Resilience refers to a supply chain's ability to foresee and evade disruptions or an ability to recuperate swiftly from failures. Carvalho et al.'s simulation model analyses possible disturbances across supply chains that could cause failures, and helps managers design risk mitigation strategies to overcome them. The model uses total cost (material, production, inventory holding and transportation) and lead time to assess the scenarios.

Nickel et al. (2012)'s multi-stochastic programming based supply chain network design framework solves the facility location problem while highlighting the risks associated with the investments in facilities. The model considers factors of return of investment and uncertainty, such as on interest rates that affect financial decisions on network design including on opening/closing a facility also, demand uncertainty, which affects extent of customer service provision. The framework uses a scenario tree to find the risk of forecasted revenue.

The framework proposed by Goh *et al.* (2007) considers all relevant risk and uncertainty aspects for a global company, and provides a solution to simultaneously minimize all of them. The multi-stage stochastic framework seeks to optimize after-tax profit by considering uncertainty factors of demand, fluctuating import tariffs, and tax charges, as well as the continually shifting exchange rates. It accordingly helps managers make decisions regarding closing/opening a facility, capacity planning and development of the distribution network.

Now that we have reviewed different supply chain design/redesign models for environmental, closed-loops, cost-efficiency and risk/uncertainty aspects; in the next phase, we will apply the combined learning from the review to understand the supply chain network issues of the case company and provide recommendations.

13.4 CASE STUDY OF THE GLOBAL TIRE MANUFACTURING FIRM

The case firm manufactures various sizes and types of tires catering to multiple industries (mining, aircraft) as well as trucks and passenger cars. Even though the firm has been expanding geographically and opening new production facilities, the current supply chain network design configuration of the firm has remained the same for the most part in the last 20 years. The focus of this case study, though, is on redesigning the supply chain network of the regional subsidiary that only handles the sales of passenger and truck tires in the Middle East and Africa (MEA) region. In the initial meeting with CEO and top management, the company's proposed strategic plan for the next five years was understood. One of the key aspects in the strategic plan was that regional customer demand must be satisfied from production within each region, as the firm will not operate any manufacturing plants in MEA; instead, this region will be integrated with Europe (EMEA), and European plants will be expected to fulfill the MEA demand in the future. In the meeting, the CEO also highlighted: (i) the need to improve customer service level (as they have long waiting times); (ii) reduce the distance between the manufacturing plants and demand points; (iii) reduce supply chain costs to remain competitive in the global environment; (iv) reduce logistics and production delays; and (v) be prepared for various supply chain risks.

13.4.1 CURRENT SUPPLY CHAIN NETWORK ANALYSIS OF THE CASE FIRM

In this section, the knowledge developed from the literature review was applied to thoroughly understand the supply chain network of the case company and delineate the supply chain issues facing the firm.

The case study's regional supply chain network contains 82 demand points located in 52 different countries in MEA with 30 corresponding manufacturing plants that are geographically dispersed and located in 12 countries.

It is to be noted that only 10 of these manufacturing plants lie under the umbrella of Europe and MEA (EMEA) subsidiaries (Europe and Turkey); therefore, the other 20 plants are supporting the region from outside of their appointed district (Japan, Thailand, Taiwan, Indonesia, and the United States). Similarly, 74% of demand in the region is met from plants outside of the region, and only 26% of the orders

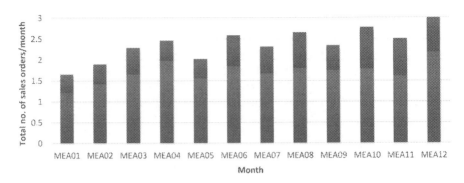

FIGURE 13.5 Monthly regional stochastic demand.

are fulfilled from production in Europe and Turkey. Of the 74%, 40% of the total orders are fulfilled by manufacturing plants in Japan (Headquarters), while plants in Thailand, Taiwan, and Indonesia fulfill more than 30% of the orders from the MEA region.

Of the total orders placed from the 52 countries in the region, 71.5% (on average) are directly shipped from manufacturing plants to demand points. The remaining 28.5% are directly shipped to Dubai (UAE), for local consolidation, and subsequently shipped to end customers in various countries, including those countries that have: (i) very low demand; (ii) political instability; and (iii) payment issues. To get deeper insights and to account for seasonality, the demand per month coming from the regional 52 countries is provided in Figure 13.5. The blue highlighted bars represent the orders that are directly shipped from the manufacturing plants, while the red shows the orders that are consolidated in the Dubai (UAE) warehouse for trans-shipment. As can be seen, the demand is seasonal; therefore, the firm is challenged with operational uncertainty in many countries.

It is clear from the current network configuration, the demand of the MEA region has imposed pressure on manufacturing plants outside of the EMEA region (to operate over capacity), while the plants within the EMEA region are underutilized. Consequently, the lead time from production to delivery for plants in Japan, Thailand, Taiwan and Indonesia is high, as these facilities are operating over capacity with associated large waiting times.

Moreover, the geographical distance from these plants to the demand points further adds to the lead time. When we analyzed the company's internal data, we identified that the average order-delivery lead times of 52 countries in the region are 30 days. However, when we analyzed at an individual country level, we found some countries with significantly higher order-delivery lead times compared to average. These countries with the highest lead times are provided in Table 13.3.

The company's goal is to reverse these numbers, which means Europe will fulfill most of the demands of the MEA region within the next five years. The capacity of the production facilities is to be assumed fixed, and that the manufacturing facilities in Europe can meet the demand of the MEA region. As mentioned earlier, in the current competitive global business environment, customers expect fast deliveries. Therefore, higher lead times mean lower quality of service, which is one of the

TABLE 13.3
Lead Time Per Days

Countries	Order-Delivery Lead Time Days
MAURITANIA	163
SUDAN	146
ALGERIA	126
COTE D'IVOIRE	96
MALTA	91
QATAR	84
KUWAIT	81
TUNISIA	77
IRAQ	72
GABON	71
NIGERIA	70

challenges the case the company is facing, and that is similar to those discussed in the literature (Fattahi *et al.*, 2018; Monostori, 2016; Stevens & Johnson, 2016). One of the critical drivers of supply chain network design is improvement of customer service levels that in turn enhances competitive advantage.

Additionally, analysis of internal data shows that some of the orders shipped directly to demand points involve small quantities. As displayed in Tables 13.4 and 13.5, countries such as Ghana and Tunisia have low demand throughout the year and orders directly shipped to them are below the company's threshold for efficient delivery, which is 1.5. The yellow highlighted data in the tables also show countries and periods with low demand and high demand variability/uncertainty; the need to consider such uncertainty in supply chain network redesign is therefore also highlighted. It was also highlighted during the interviews that profit margins for passenger car tires are low; therefore, any inefficiency in the design of the logistics network has a significant bearing on business profitability.

TABLE 13.4
Shipment Below the Efficiency Level (Highlighted in Yellow)

Month	Countries	Japan	Indonesia	Thailand
Mar	GHANA		0.38	
May	GHANA		0.75	
Jun	GHANA	0.87		
Jul	GHANA	0.44		
Aug	GHANA	0.68		0.59
Dec	GHANA			0.89

TABLE 13.5
Evidence of Non-Optimal Ordering (Highlighted in Yellow)

Month	Countries	Japan	Thailand	Europe
Jan	TUNISIA	0.59		
Apr	TUNISIA	0.19		
May	TUNISIA		0.17	
Jun	TUNISIA			0.71
Sep	TUNISIA			0.32
Nov	TUNISIA			0.59

The manufacturing locations are also not optimally allocated to demand points; the lower demand locations are allocated to over capacitated plants that provide slower deliveries (refer to Tunisia in the Table 13.5).

The low quantity and (still) regular/frequent shipments have raised the question of whether increasing the capacity of the consolidation point in UAE by extending the facility and consolidating smaller orders is the optimal solution. For example, in that case, orders for Ghana for June, July and August can be shipped from Japan to Dubai, and consolidated and shipped to Ghana all at once in one shipment. Facility location problems, closure and opening have been addressed in many of the network design models discussed in the literature. Inventory Optimization Model found in the literature could be the most appropriate model for handling inventory operations since the model works by considering lead-time subject to demand and forecasting uncertainty.

Another critical challenge that needs to be factored in is the political and financial instability in the region, which results in fluctuating exchange rates, as well as a change in customs regulations. As was seen in the literature, strategic uncertainty is another driver of supply chain network design. As per internal documents and discussions with executives, for countries such as Ethiopia and Algeria, the organization is facing significant challenges in terms of credit payments as a result of unbalanced exchange rates.

The other important issue facing the company is the lack of consideration for environmental issues while meeting the demand for the 52 countries in the region. This is concerning given that the case company global headquarters insists considerations for environmental aspects in the supply chain, including a reduction in CO_2 emissions, specifically, reduction of at least 50% GHG emissions worldwide by 2050. Further, there is no recycling program for the development of closed-loop supply chains in the case company. Again, the global headquarters insists that for every new tire the company sells, one company tire or any one tire needs to be recovered and put to reuse or recycle. There is pressure on the case company to implement this.

Overall, the analysis of the case company demonstrated that the firm is facing many of the same challenges that were discussed in the literature, such as inefficiencies in the supply chain network along with strategic and operational uncertainty due to political and demand uncertainty. Furthermore, longer lead times, reduce the company's competitive advantage and customer service level, which were another two reasons highlighted for network design; lastly, frequent low order quantity shipments are not cost-effective. Therefore, it can be stated that the company will benefit from implementing supply chain network design configuration, which will optimize the network and potentially reduce their associated logistics network cost. Moreover, redesigning the supply chain network with strategies that can withstand any levels of uncertainty, will increase the case company's supply chain network resilience.

13.4.2 RECOMMENDATIONS FOR THE CASE FIRM

In this section, the understanding from the literature is used to propose recommendations that address the issues identified for the case study. As discussed in the literature, configuring a supply chain network for any company is an intricate process; it is even more so for the case company as its network is very complex. Its supply chain network design should involve an integrated strategic and tactical level decision of network design models, as the supply chain network design decisions that involve facility location and capacity allocation are part of the long-term strategic decisions. In comparison, reconfiguration of the distribution network is a tactical level issue, which requires mid-term planning.

Furthermore, the case firm's supply chain network design problem is far more complex than any of the models we have come across in the literature, since the case firm's network redesign has multiple objectives which include optimal facility location for each demand point, and uncertainty of demand and political instability, therefore a decomposition method is recommended. Decomposition approaches allow managers to make realistic and detailed decisions at each of the levels of the supply chain network (Dullaert *et al.*, 2007). The company should segregate the countries based on the challenges they face in each.

It is recommended that the case company consider implementing optimization network design models for countries that have minimum levels of uncertainty. Optimization supply chain network design models with certain demand quantity of Margolis *et al.* (2018) and Hammami and Frein (2014) that provide solutions for facility location problems, and the design of the distribution network can be adopted. These models also address facility opening problems. The network optimization design models of Zheng *et al.* (2019) that address lead time constraints with certain demand could be used to optimize lead time for those countries that are facing longer order-delivery time but have certain demand. On the other hand, those locations which have issues with demand uncertainty should be redesigned with models that address uncertainty (Jahani *et al.*, 2018; Rahimi *et al.*, 2019).

The supply chain network of the countries that are faced with stochastic demand and strategic level of uncertainty could be configured with the frameworks that address uncertainty such as the model proposed by Fattahi *et al.* (2018). These models could be implemented to mitigate the uncertainty of demand while redesigning the supply chain network; most of these models use stochastic programming and scenario analysis to mitigate the risk while solving the production-distribution problem. Supply chain network design models such as Rahimi *et al.* (2019) and Goh *et al.* (2007) could be implemented for countries that have both demand and political uncertainty. These models address the optimal selection of distribution links and facility locations (including inventory and warehouse) with consideration of risks.

For integrating environmental issues, a multi-objective linear programming model similar to the one proposed by Chanchaichujit *et al.* (2020) for the tire industry could be used to optimize total costs and total GHG emissions simultaneously. The total costs can include all supply chain costs, such as production, inventory and transportation.

Furthermore, for the circular economy, the supply chain network design of the retreading plants and collection points for old tires could be executed with the use of the models of Amin *et al.* (2017) and Pedram *et al.* (2017).

13.5 CONCLUSION AND LESSONS LEARNED

In this paper, the objectives have been met, as many supply chain network design models were reviewed to provide managerial support for the case firm's network design planning. This includes a review of the models and approaches that optimize supply chain networks by reducing the cost of operation or maximizing profit while focusing on improving customer service level (lead time). Similarly, levels of uncertainty were first identified, followed by the introduction of models that configure supply chain networks with solution approaches to multiple levels of uncertainty. Environmental and closed-loop models were analyzed to support the supply chain network design of the case firm. Importantly, the understanding garnered from the review was helpful in first understanding the network issues of the case firm and then provide valuable recommendations.

Given the valuable insights synthesized from the review and case study, the literature and case study be understand the network issues to any firm in any industry. The studies discoursed in this paper could provide implications for practitioners that are considering supply chain network design by guiding them to the right network design models based on their supply chain network design objectives. Similarly, through the analysis and review of the recent works of literature, this paper has added to the body of the supply chain network design models by highlighting the gaps that exist in literature. Therefore, supply chain network academics may find this paper helpful to recognize the gaps in the models and consequently build on the knowledge.

The study has some limitations. While the literature review for this paper was extensive, it is not exhaustive or systematic. In future research, additional aspects can be considered in the supply chain network design/redesign such as customs, duties, and the clearance time at ports. Despite the limitations, we think that the findings of this study can significantly contribute towards the advancement of supply chain network design/redesign models and encourage more research in this field.

REFERENCES

Amin, S. H., Zhang, G., & Akhtar, P. (2017). Effects of uncertainty on a tire closed-loop supply chain network. *Expert Systems with Applications, 73*, 82–91. doi: 10.1016/j. eswa.2016.12.024.

Ansari, Z. N., & Kant, R. (2017). Exploring the framework development status for sustainability in supply chain management: A systematic literature synthesis and future research directions. *Business Strategy and the Environment, 26*(7), 873–892.

Buddadee, B., Wirojanagud, W., Watts, D. J., & Pitakaso, R. (2008). The development of multi-objective optimization model for excess bagasse utilization: A case study for Thailand, *Environmental Impact Assessment Review, 28*(6), 380–391.

Carvalho, H., Barroso, A. P., Machado, V. H., Azevedo, S., & Cruz-Machado, V. (2012). Supply chain redesign for resilience using simulation. *Computers & Industrial Engineering, 62*(1), 329–341. doi: 10.1016/j.cie.2011.10.003.

Chanchaichujit, J., Saavedra-Rosas, J., Quaddus, M., & West, M. (2016). The use of an optimisation model to design a green supply chain: A case study of the Thai rubber industry. *The International Journal of Logistics Management, 27*(2), 595–618. https://doi.org/doi:10.1108/IJLM-10-2013-0121

Chanchaichujit, J., Balasubramanian, S., & Shuka, V. (2020). Multi-objective decision model for green supply chain management. *Cogent Business & Management, 7*(1), 1783177.

Coyle, J. J., Bardi, E. J., & Langley, C. J. (2004). *The management of business logistics: A supply chain perspective.* Mason: South-Western/Thomson Learning

Creazza, A., Dallari, F., & Rossi, T. (2012). Applying an integrated logistics network design and optimisation model: The Pirelli Tyre case. *International Journal of Production Research, 50*(11), 3021–3038. doi: 10.1080/00207543.2011.588614.

Croxton, K. L., & Zinn, W. (2005). Inventory considerations in network design. *Journal of Business Logistics, 26*(1), 149–168.

Dullaert, W., Bräysy, O., Goetschalckx, M., & Raa, B. (2007). Supply chain (re) design: Support for managerial and policy decisions. *European Journal of Transport and Infrastructure Research, 7*(2).

EIA-IEO (2013). International Energy Outlook 2013. U.S. Energy Information Administration, Retrieved March 15, 2015, from https://www.eia.gov/outlooks/ieo/pdf/0484(2013).pdf

EPA (2017). Climate Change Indicators: Sea Level. United States Environmental Protection Agency, Retrieved March 12, 2017, from https://www.epa.gov/climate-indicators/climate-change-indicators-sea-level

Fattahi, M., Govindan, K., & Keyvanshokooh, E. (2018). A multi-stage stochastic program for supply chain network redesign problem with price-dependent uncertain demands. *Computers & Operations Research, 100*, 314–332. doi: 10.1016/j.cor.2017.12.016.

Garcia, D. J., & You, F. (2015). Supply chain design and optimization: Challenges and opportunities. *Computers & Chemical Engineering, 81*, 153–170. doi: 10.1016/j. compchemeng.2015.03.015.

Goh, M., Lim, J. Y. S., & Meng, F. (2007). A stochastic model for risk management in global supply chain networks. *European Journal of Operational Research, 182*(1), 164–173. doi: 10.1016/j.ejor.2006.08.028.

Graves, S. C., & Willems, S. P. (2000). Optimizing strategic safety stock placement in supply chains. *Manufacturing & Service Operations Management, 2*(1), 68–83.

Habib, M. A., Bao, Y., & Ilmudeen, A. (2020). The impact of green entrepreneurial orientation, market orientation and green supply chain management practices on sustainable firm performance. *Cogent Business & Management, 7*(1), 1743616.

Hammami, R., & Frein, Y. (2014). Redesign of global supply chains with integration of transfer pricing: Mathematical modeling and managerial insights. *International Journal of Production Economics, 158*, 267–277. doi: 10.1016/j.ijpe.2014.08.005.

Hammami, R., Frein, Y., & Hadj-Alouane, A. B. (2008). Supply chain design in the delocalization context: Relevant features and new modeling tendencies. *International Journal of Production Economics*, *113*(2), 641–656. doi: 10.1016/j.ijpe.2007.10.016.

Hugo, A., & Pistikopoulos, E. N. (2005). Environmentally conscious long-range planning and design of supply chain networks. *Journal of Cleaner Production*, *13*(15), 1471–1491. https://doi.org/10.1016/j.jclepro.2005.04. 011

IPCC (2007). Climate Change 2007: Synthesis Report. Contribution of Working Groups I, II and III to the Fourth Assessment Report of the Intergovernmental Panel on Climate Change, Retrieved March 12, 2017, from https://www.ipcc.ch/pdf/assessment-report/ar4/syr/ar4_syr_full_report.pdf

Jaehne, D. M., Li, M., Riedel, R., & Mueller, E. (2009). Configuring and operating global production networks. *International Journal of Production Research*, *47*(8), 2013–2030. doi: 10.1080/00207540802375697.

Jahani, H., Abbasi, B., Alavifard, F., & Talluri, S. (2018). Supply chain network redesign with demand and price uncertainty. *International Journal of Production Economics*, *205*, 287–312. doi: 10.1016/j.ijpe.2018.08.022.

Jawjit, W., Pavasant, P., & Kroeze, C. (2015). Evaluating environmental performance of concentrated latex production in Thailand. *Journal of Cleaner Production*, *98*, 84–91.

Kenneth Research (2019). Industrial rubber market share, trend, opportunity and forecast. Kenneth Research. https://www.kennethresearch.com/report-details/industrial-rubber-market/10075809

Kim, N. S., Janic, M., and & Wee, B. (2009). Trade-off between carbon dioxide emissions and logistics costs based on multiobjective optimization. *Transportation Research Record*, *2139*(1), 107–116.

Krungsri Report (2019). Natural rubber processing: Thailand industry outlook 2019–2021. Krungsri: Bank of Ayudha.

Lockett, H., Raval, A., & Sheppard, D. (2019). Oil prices soar after attacks halve Saudi output, Financial Times. Available at: https://www.ft.com/content/353bce38-d806-11e9-8f9b-77216ebe1f17 (*Accessed 16 September 2019*).

Margolis, J. T., Sullivan, K. M., Mason, S. J., & Magagnotti, M. (2018). A multi-objective optimization model for designing resilient supply chain networks. *International Journal of Production Economics*, *204*, 174–185. doi: 10.1016/j.ijpe.2018.06.008.

Mokhtar, A. R. M., Genovese, A., Brint, A., & Kumar, N. (2019). Supply chain leadership: A systematic literature review and a research agenda. *International Journal of Production Economics*, *216*, 255–273. doi: 10.1016/j.ijpe.2019.04.001.

Monostori, J. (2016). Robustness- and complexity-oriented characterization of supply networks' structures. *Procedia CIRP*, *57*, 67–72. doi: 10.1016/j.procir.2016.11.013.

Nickel, S., Saldanha-da-Gama, F., & Ziegler, H. P. (2012). A multi-stage stochastic supply network design problem with financial decisions and risk management. *Omega*, *40*(5), 511–524. doi: 10.1016/j.omega.2011.09.006.

Pedram, A., Yusoff, N. B., Udoncy, O. E., Mahat, A. B., Pedram, P., & Babalola, A. (2017). Integrated forward and reverse supply chain: A tire case study. *Waste Management*, *60*, 460–470. doi: 10.1016/j.wasman.2016.06.029.

Rahimi, M., Ghezavati, V., & Asadi, F. (2019). A stochastic risk-averse sustainable supply chain network design problem with quantity discount considering multiple sources of uncertainty. *Computers & Industrial Engineering*, *130*, 430–449. doi: 10.1016/j.cie.2019.02.037.

Srivastava, S. K. (2007). Green supply-chain management: A state-of-the-art literature review. *International Journal of Management Reviews*, *9*(1), 53–80.

Stevens, G. C., &Johnson, M. (2016). Integrating the supply chain ... 25 years on. *International Journal of Physical Distribution & Logistics Management*, *46*(1), 19–42. doi: 10.1108/IJPDLM-07-2015-0175.

Subulan, K., Taşan, A. S., & Baykasoğlu, A. (2015). Designing an environmentally conscious tire closed-loop supply chain network with multiple recovery options using interactive fuzzy goal programming. *Applied Mathematical Modelling, 39*(9), 2661–2702. doi: 10.1016/j.apm.2014.11.004.

UNEP-EGR (2014). The Emissions Gap Report 2014. United Nations Environment Programme, Nairobi, Retrieved April 07, 2015, from http://www.unep.org/emissionsgapreport2014

UNEP-EGR (2016). The Emissions Gap Report 2016. United Nations Environment Programme, Nairobi, Retrieved March 12, 2017, from http://uneplive.unep.org/theme/index/13#egr

Vaitheesvaran, B. (2019). Made-in-India iPhone X from July. [online] The Economic Times. Available at: https://economictimes.indiatimes.com/tech/hardware/made-in-india-iphone-x-from-july/articleshow/68824180.cms *(Accessed 5 September 2019).*

Varzandeh, J., Farahbod, K., & Zhu, J. (2016). Global logistics and supply chain risk management. *Journal of Business and Behavioral Sciences, 28*(1), 124–130.

Wang, F., Lai, X., & Shi, N. (2011). A multi-objective optimization for green supply chain network design. *Decision Support Systems, 51*(2), 262–269.

Winebrake, J. J., Corbett, J. J., Falzarano, A., Hawker, J. S., Korfmacher, K., Ketha, S., & Zilora, S. (2008). Assessing energy, environmental, and economic tradeoffs in intermodal freight transportation. *Journal of the Air & Waste Management Association, 58*(8), 1004–1013.

You, F., & Wang, B. (2011). Life cycle optimization of biomass-to-liquid supply chains with distributed–centralized processing networks. *Industrial & Engineering Chemistry Research, 50*(17), 10102–10127.

Zheng, X., Yin, M., & Zhang, Y. (2019). Integrated optimization of location, inventory and routing in supply chain network design. *Transportation Research Part B: Methodological, 121*, 1–20. doi: 10.1016/j.trb.2019.01.003.

Index

Printed in the United States
By Bookmasters